Natural Dyes for Textiles

THE TEXTILE INSTITUTE BOOK SERIES

Incorporated by Royal Charter in 1925, The Textile Institute was established as the professional body for the textile industry to provide support to businesses, practitioners and academics involved with textiles and to provide routes to professional qualifications through which Institute Members can demonstrate their professional competence. The Institute's aim is to encourage learning, recognise achievement, reward excellence and disseminate information about the textiles, clothing and footwear industries and the associated science, design and technology; it has a global reach with individual and corporate members in over 80 countries.

The Textile Institute Book Series supersedes the former 'Woodhead Publishing Series in Textiles', and represents a collaboration between The Textile Institute and Elsevier aimed at ensuring that Institute Members and the textile industry continue to have access to high calibre titles on textile science and technology.

Books published in The Textile Institute Book Series are offered on the Elsevier web site at: www.elsevier.com/books-and-journals and are available to Textile Institute Members at a substantial discount. Textile Institute books still in print are also available directly from the Institute's web site at: www.textileinstitute.org

To place an order, or if you are interested in writing a book for this series, please contact Matthew Deans, Senior Publisher: m.deans@elsevier.com

Recently Published and Upcoming Titles in The Textile Institute Book Series

The Textile Institute Book Series

Natural Dyes for Textiles

Sources, Chemistry and Applications

Padma Shree Vankar

The Textile Institute

WOODHEAD
PUBLISHING
An imprint of Elsevier

Woodhead Publishing is an imprint of Elsevier
The Officers' Mess Business Centre, Royston Road, Duxford, CB22 4QH, United Kingdom
50 Hampshire Street, 5th Floor, Cambridge, MA 02139, United States
The Boulevard, Langford Lane, Kidlington, OX5 1GB, United Kingdom

Library of Congress Cataloging-in-Publication Data
A catalog record for this book is available from the Library of Congress

British Library Cataloguing-in-Publication Data
A catalogue record for this book is available from the British Library

ISBN: 978-0-08-101274-1 (print)
ISBN: 978-0-08-101884-2 (online)

For information on all Woodhead publications visit our
website at https://www.elsevier.com/books-and-journals

  Working together
to grow libraries in
developing countries

www.elsevier.com • www.bookaid.org

Publisher: Matthew Deans
Acquisition Editor: David Jackson
Editorial Project Manager: Edward Payne
Senior Production Project Manager: Priya Kumaraguruparan
Cover Designer: Vicky Pearson Esser

Typeset by SPi Global, India

Contents

Newer natural dyes for cotton   1A

P.S. Vankar
FEAT (Facility for Ecological and Analytical Testing), Kanpur Kalyanpur, India

Introduction

In India with its diversity of climatic zones and altitudes, geographic variety has resulted in a rich biodiversity that has gifted flora and fauna, which yields an array of dye-producing shrubs and perennials. The ancient people exclusively used dyestuffs of vegetables, minerals, and animal origin, all easily obtained in their own vicinity. In India, it was widely used for coloring of fabrics and other materials. In order to meet the growing demand for natural colorants, new pigment crops are being sought. The newer sources need to be tapped. However, for some sources such as Al root (*Morinda tinctoria*), manjistha, safflower, and indigo, there is an organized cultivation because of their exceptional and established versatility. Increased acreage is devoted to Al root and safflower cultivation purely because of their dye qualities. More plants need to be considered as agronomically viable plantations.

Natural dyes are basically elements of natural resources, and these dyes are generally classified as plant, animal, mineral, and microbial dyes based on their source of origin, although plants are the major sources of natural dyes. Natural dyes per se are sustainable as they are renewable and biodegradable. The experience with natural dyeing has given an insight to explore plants in the neighborhood. Finding fiber colors in plants that grow easily and fast has lead into a new world of fiber colors that give exotic shades. These natural colors have richness and luster that synthetics can never attain. It has become a common misconception that natural dyes only produce beiges and browns and other washed out shades. In reality, vibrant, fast natural colors can be produced, which are comparable with and often surpass the colors of synthetics. Apart from the sources of these dyes, it is perhaps the commitment of those propagating them and the near clinical efficiency with which dye is extracted, produced, and used, which is responsible for the unique nature of natural dyeing and producing stable coloration.

Experimental trials were carried out in domestic gardens in collaboration with botanists mainly focusing on the best conditions for the growth of dye plants in regard to soil and climatic factors. Modern cultivation system for getting maximal dye yields including optimal seeding and harvesting time and optimal fertilization procedures was adapted. The utilizable plant parts were subjected to specific dehydration processes, or the dyestuff was extracted as per the given strategy.

It will help in giving an idea about feasibility of providing high-quality natural dyes from plants, creating new opportunities for both farmers and fabric industry—in line with the current consumer trends toward ethnic fabric and natural products. Several newer sources of natural dyes particularly for dyeing cotton and silk fabrics to get a gamut of colors have been explored. This will increase the availability of newer shades and new choices of fabric dyed with natural dyes.

Natural Dyes for Textiles. http://dx.doi.org/10.1016/B978-0-08-101274-1.00001-X

The aim is to show feasibility of providing high-quality natural dyes from plants, creating new opportunities for both farmers and the fabric industry—in line with the current consumer trends toward ethnic fabric and natural eco-friendly products.

Many new plant sources have been explored for natural dyeing, and the following section is a brief introduction of the plants used from our screening experience. Many of them are known medicinal plants and have been shown to have potentially rich natural colorant content as well. It is imperative to make a documentation of these plants for the future use of these plants as sources of natural dyes.

1A.1 *Acacia Arabica* bark

Plant: *Acacia arabica*
Family: Fabaceae
Genus: *Acacia*
Part used: bark

Fig. 1A.1 Acacia bark.

Acacia arabica (acacia bark) belongs to the class Dicotyledonae of family Fabaceae and subfamily Mimosoideae (Fig. 1A.1). It is primarily used for fuel wood, timber, agroforestry, and vegetable tanning. The bark powder is astringent in nature, and there is only one report where acacia has been used as dye in Nigeria. Gum arabic is a by-product of this tree, which is used in calico printing and dyeing. *Acacia arabica* is a perennial shrub or tree, 2.5–10 (–20) m tall, variable in many aspects.[1] The branches spread in a way forming a dense flat or rounded crown with dark- to black-colored stems; the bark is thin, rough, fissured, and deep red-brown in color having spines (thorns) thin and straight, however, mature trees are commonly without thorns. The bark of the plant was used for dyeing for cotton on the basis of dye content from the bark (Vankar, 2002; Tiwari et al., 2001).

[1] http://database.prota.org/PROTAhtml/Acacia%20nilotica_En.htm

The bark powder of the plant acacia was heated with water, and the color was extracted and used for dyeing (Mohanty et al., 1987).

Natural dye from babul bark powder is safe and eco-friendly as it has been shown to be free from hazardous chemicals. Therefore, its commercial use shall definitely minimize the health hazards caused by the use of synthetic dyes. The plant was known as food and dye source to the Central Mexican population in the 14–16th centuries. However, use of the extract of the bark for natural dyeing of cotton fabric has been revived.

1A.2 *Mahonia napaulensis* DC

Plant: *Mahonia napaulensis*
Family: Berberidaceae
Genus: *Mahonia*
Part used: stem

Mahonia napaulensis DC (local name Taming) family Berberidaceae contains a natural cationic colorant. *Mahonia napaulensis* is an evergreen shrub growing to 2.5 m (8 ft) by 3 m (9 ft).[2] It is in leaf till January. It is in flower from March to April. The flowers are hermaphrodite (have both male and female organs) and are pollinated by insects. Suitable for light (sandy), medium (loamy), and heavy (clay) soils and can grow in heavy clay soil. Suitable pH are acid, neutral, and basic (alkaline) soils. It can grow in semishade (light woodland) or no shade. It prefers moist soil. This dye has been used for dyeing textiles by the Apatanis (a tribe of Arunachal Pradesh) since ancient times. Dyeing with *M. napaulensis* with different pretreated cotton showed marked improvement in dye uptake. The pretreatments provided improved exhaustion of colorant from *Mahonia* stem extract (Fig. 1A.2).

Fig. 1A.2 Mahonia stem.

[2] http://www.pfaf.org/user/Plant.aspx?LatinName=Mahonia+napaulensis

In the quest to improve the dye adherence of the natural dye available from mahonia, several pretreatments such as cellulase, sodium lauryl sulfate, and ammonium sulfamate were used in order to improve the dye adherence to cotton fabric. *Mahonia* grows abundantly in the Ziro valley of Apatani plateau in Lower Subansiri district. This indigenous dye can be produced in large scale commercially.

1A.2.1 Traditional knowledge of natural dyeing

The Apatani tribe of Arunachal Pradesh has been engaged in extraction, processing, and preparation of dyes using stem and root bark of *M. napaulensis* plant (Mahanta and Tiwari, 2005; Usher, 1974; Vankar et al., 2008a). The practice of indigenous systems for preparing dyestuffs and the processes of dyeing can be developed using modern technological methods. The work on *Mahonia* was designed with an aim to focus on the innovative methods of dye extraction, mordant study, and by means of application of modern technology to sharpen the skill of traditional tribal Arunachalee dyers.

Ammonium sulfamate with *M. napaulensis* stem extract was found to enhance the dyeability and fastness properties.

1A.3 Salvia splendens (Fig. 1A.3)

Plant: *Salvia splendens*
Family: Lamiaceae
Genus: *Salvia*
Part used: flower (red)

Fig. 1A.3 Salvia flower.

Plant taxonomy classifies red salvia plants as *S. splendens*, commonly referred as scarlet sage; many people refer to the plants simply as red salvia, indigenous to Brazil, but it is also found in India as seasonal flowering plant. Red salvia flowers are known as annual plants in temperate zones: they are damaged by hard frosts and will not survive through cold winters.

Salvia splendens (scarlet sage and tropical sage) is a tender herbaceous perennial native to Brazil; however, it also grows in India, growing at an altitude of 2000–3000 m above the sea level, in a humid and warm area.[3] The plant reaches 1.3 m (4.3 ft) height. Salvia prefers moist soil and well-drained soil with full sunlight for its healthy growth.

Salvia is a tender tropical plant that is typically grown as a warm weather annual bedding plant. It has long been a garden standard, reliably blooming over an extended period. Ever more varieties are being developed, giving a wide range of colors, including white, salmon, and purple, as well as the traditional bright red, and heights from about 8 in. (20 cm) to nearly 3 ft (0.9 m). It is the bright red variety that was used for the dyeing purpose. Leaves are bright to dark green, elliptical, and toothed. Flowers grow on spikes and are two-lipped, with a flat lower lip and helmet-shaped upper lip.

Salvias are prolific and durable bloomers. They are not bothered by pests. Unfortunately, the selected varieties of this species are not as attractive to hummingbirds.

Although the scarlet varieties are the best known, *S. splendens* does come in other colors. Dark red variety of salvia flowers were collected from the forest of Tawang, Arunachal Pradesh, and were identified by horticulture department.

Conventional dyeing properties of cotton with the aqueous extract of salvia flowers (red variety) (Vankar and Kushwaha, 2011) by using tin mordant. CIELab values, K/S values, and fastness properties of the dyed fabrics were ascertained. The dye showed promising results and acceptability for commercial dyeing. Tin mordant was chosen particularly as it gave stability to the extract and the color content deepened.

1A.4   *Canna indica* (Fig. 1A.4)

Plant: *Canna indica*
Family: Cannaceae
Genus: *Canna*
Part used: flower (red)

Canna indica is a rhizomatous, perennial, erect, robust herb, up to 3.5 m tall. Rhizome is branching horizontally, up to 60 cm long and 10 cm in diameter, with fleshy segments resembling corms, covered with scale leaves, and thick fibrous root.[4] Fleshy stem, arising from the rhizome, is usually 1–1.5 m tall, often tinged with purple. Leaves arranged spirally with large open sheaths, sometimes shortly

---

[3] https://en.wikipedia.org/wiki/Salvia_splendens
[4] https://en.wikipedia.org/wiki/Canna_plant

Fig. 1A.4 Red canna flower.

petiolate; blade narrowly ovate to narrowly elliptical, up to 60 cm × 15–27 cm, entire, base rounded to cuneate, gradually attenuate to the sheath, apex acuminate, midrib prominent, underside often slightly purplish. Inflorescence terminal, racemose, usually simple but sometimes branched, bearing single or paired irregular, bisexual flowers. *Canna indica* grows well in various tropical and subtropical climates. It seems to be day length neutral, as it grows and flowers under a broad range of photoperiodic conditions. It is affected by drought but tolerates excessive moisture (but not water logging). It is very tolerant of shade. Normal growth occurs at temperatures above 10°C, but it also survives high temperatures of 30–32°C. Normal growth occurs at temperatures above 10°C, but it also survives high temperatures of 30–32°C and tolerates light frost. *Canna indica*'s preferred soils are deep sandy loams, rich in humus. It tolerates a pH range of 4.5–8.0. This hybrid variety gives ample of flowers yielding good amount of natural colorant. The other advantage of the canna plant is that it is self-propagating.

The ease of solubilization of canna dye in ethanol and its influence in dye exhaustion and fixation on cotton fabric was found to be the best studied in details (Srivastava et al., 2008).

Earlier report from our laboratory showed the use of aqueous extract of canna dye for cotton fabric using eco-friendly mordants (Ghorpade et al., 2000), but during the process, it was found that a major part of the colorant goes as a waste. Thus, in order to improve the color strength, a method for better solubilization of this partially soluble dye was devised, and it is interesting to note that microemulsion systems can be used for solubilization of such dyes very efficiently and effectively.

The water insoluble natural dyes can create problems during their application, leading to nonuniform and uneven dyeing. Canna dye shows poor solubility in water. Hence, attempts were made for dissolution of partially insoluble natural canna dye in different mediums. Aqueous, methanolic/ethanolic and oil/water microemulsion were the solvent systems used for this plant. The ease of solubilization of canna dye in ethanol and its influence in dye exhaustion and fixation on cotton fabric was found to be the best.

1A.5 *Rhododendron arboreum*

Plant: *Rhododendron arboreum*
Family: Ericaceae
Genus: *Rhododendron*
Part used: flower (red)

Fig. 1A.5 Rhododendron flowers.

Rhododendron arboreum, the tree rhododendron, is an evergreen shrub or small tree with a showy display of bright red flowers (Fig. 1A.5). It is found in Bhutan, China, India, Myanmar, Nepal, Sri Lanka, and Thailand. *Rhododendron arboreum* is the national flower of Nepal; in India, it is the state tree of Uttarakhand and state flower of Himachal Pradesh and Nagaland. It prefers moist but well-drained, leafy, humus-rich, acid soil, in dappled shade. It has broad, dark green leaves, 7–19 cm (3–7 in.) long, with a silvery, fawn, or brown hairy coating beneath.[5] In early and midspring, trusses of 15–20 bell-shaped flowers are produced in red, pink, or white. They have black nectar pouches and black spots inside. Red variety of rhododendron has been used for dyeing purpose.

Rhododendron arboreum is one of the most stately and impressive rhododendron species. It is extremely variable in stature, hardness, and flower color and leaf characteristics. Its species name arboreum means treelike. Originally discovered in north-central India, the plant is found in the Himalayas from Kashmir to Bhutan and on the hills of Assam and Manipur. It grows at elevations of 4500–10,500 ft and grows up

[5] https://en.wikipedia.org/wiki/Rhododendron_arboreum

to 40–50 ft high sometimes attaining over 100 ft. This is an evergreen much branched tree growing up to 1 m in height and 2.4 m in width. Flowering season is from March to April/June to September bearing deep red or crimson to pale pink flowers. Bearing up to 20 blossoms in a single truss, this rhododendron is a spectacular sight when in full bloom. It is reported that the bright red forms of this rhododendron are generally found at the lower elevations.

The plant prefers light (sandy) to medium (loamy) soil and requires fairly moist and acidic soil. It can grow in semishade (light woodland) or no shade, requires protection from hot afternoon sun and thus requires a place in the green house or conservatory. The aesthetically appealing flowers owe its religious significance; it is considered sacred and offered in temples for ornamenting purposes.

Rhododendron arboreum is important plant of hilly region with good amount of commercial uses. The plant exhibited anti-inflammatory, hepatoprotective, antidiarrheal, antidiabetic, and antioxidant properties due to the presence of flavonoids, saponins, tannins, and other phytochemicals. Fresh petals are processed to prepare subacid jelly and sherbet a famous market commodity.

Young leaves are poisonous that cause intoxication in large quantities. The plant has a special place in the cultural and economic life of the people. It is offered in temples and religious places. The plant has good amount of medicinal and ornamental values.

Rhododendron flower extract has been studied by some researchers—Zhu and Sibiao (1991) studied the effect of metal ions on red pigments obtained from Mulberry fruit and rhododendron flower, as both contained red pigments. These red pigments were identified as anthocyanidin compounds and could be used as a red food dye. Fe^{3+} and Sn^{2+} changed the color of the red pigment from rhododendron flower, Al^{3+} and Zn^{2+} intensified the redness, and Cu^{2+} and Ca^{2+} had no effect. Conventional dyeing of cotton fabric with the aqueous extract of rhododendron flowers (red variety) by using metal mordants—$FeSO_4$, $SnCl_2$, $CuSO_4$, $SnCl_4$, $K_2Cr_2O_7$, and alum. CIE Lab values, K/S values, and fastness properties of the dyed fabrics were ascertained. The dye showed promising results and acceptability for commercial dyeing (Vankar and Shanker, 2010).

1A.6   *Cosmos sulphureus*

Plant: *Cosmos sulphureus*
Family: Asteraceae
Genus: *Cosmos*
Part used: flower (mustard color)

Cosmos[6] is a fast-growing and free-flowering annual (Fig. 1A.6). Cosmos is a tender annual that grows easily from seed and will thrive in any average, well-drained soil. Full sun dry soil and nutrition grow the best plants and will help avoid diseases. The plant has green foliage and blooms of white, pink, red, and orange.

[6] http://localcolordyes.com/2012/09/16/orange-cosmos-cosmos-sulphureus/

Fig. 1A.6 Cosmos flower.

The orange-yellow toned *C. sulphureus* variety, however, is a tender flower, but it is easy to grow by seeding directly. The cosmos flowers are harvested frequently, two to three times a week, and simply frozen in plastic bags as we picked them, thus collecting a lot over time. Cosmos produces many flowers all summer.

In the search for newer flora dyeing cotton from natural colorants, *C. sulphureus* was used for study. Both alcoholic extract and aqueous extracts were used for dyeing. There has been one report on wool dyeing with aqueous extract of cosmos flower petals (Kale et al., 2006).

Mustard orange cosmos are pH sensitive and gets redder colors if the pH is high. After straining out the flowers, ammonia is used to increase the pH. In the initial dye baths, ammonia is added before adding the wool yarn and premordanted with aluminum sulfate.[7]

In the search for newer flora dyeing cotton from natural colorants, *C. sulphureus* was used for the present study. Both alcoholic extract and aqueous extracts were used for dyeing. Mordants used were alum, stannic chloride, stannous chloride, and ferrous sulfate. Sonicator and microwave dyeing methods were tried and compared with conventional dyeing. Dye uptake was better in methanolic extract dyeing, while among the dyeing mode, microwave showed faster dye uptake.

1A.7   *Terminalia arjuna*

Plant: *Terminalia arjuna*
Family: Combretaceae
Genus: *Terminalia*
Part used: bark

[7] http://tauton.com

Fig. 1A.7 *Terminalia arjuna* (arjuna).

Terminalia arjuna (arjuna) is a genus that consists of large hard-wooded trees (Fig. 1A.7). Over 100 species widely distributed in India, *T. chebula*, *T. bellirica*, and *T. ciliata* are major related species, commonly found in almost every part of India. Every part of *Terminalia* has useful medicinal properties.[8] Arjuna holds a reputed position in both Ayurvedic and Yunani systems of medicine. It is used as expectorant, aphrodisiac, tonic, and diuretic. Standardization and optimization of dye extraction were carried out for *T. arjuna* bark. The debility of the aqueous extract was evaluated for dyeing cotton fabric. Dyed cotton fabric showed good fastness properties. It can be considered commercially viable natural dye source. This can be a good source of natural dye on the basis of their low cost, easily availability, bulk isolation, and their technoeconomics. *Terminalia arjuna* is a large tree with drooping branches and spreading evergreen all throughout the year. It grows up to 25 m height or even taller and has its bark gray and smooth. Every part of the *Terminalia* tree has useful medicinal properties. According to Ayurveda, it is alexiteric, styptic, tonic, anthelmintic, and useful in fractures, ulcers, heart diseases, biliousness, urinary infection, asthma, tumors, leucoderma, anemia, excessive perspiration, and many other medical conditions. According to Yunani system of medicine, it is used both externally and internally in urinary tract infections.

The arjun tree is recommended for reclamation of saline, alkaline soils, and deep ravines. It is used for agro- and social forestry. The timber is used for carts, agricultural implements, water troughs, traps, boat building, house building, electric poles, tool handles, jetty piles, and plywood. Fodder is useful for tasar silkworm. It is one of the major tannin-yielding trees. Bark (22%–24%), leaf (10%–11%), and fruit (7%–20%) contain tannin.[9]

[8] http://www.planetayurveda.com/library/arjuna-terminalia-arjuna
[9] http://naturalhomeremedies.co/Arjuna.html

Chemical constituents: a glucoside—arjunetin—has been isolated from bark. Recently, new flavone—arjunone—has been isolated from fruits along with cerasidin, β-sitosterol, friedelin, methyl oleanolate, gallic, ellagic, and arjunic acid.[10]

Brown color was extracted from the bark of *T. arjuna*. Bark of *T. arjuna* is flat or slightly curved, external surface pink or flesh colored with a mealy coating, and inner surface reddish brown, finely striated, peeling out in thin flakes, odorless, gritty, and astringent. The pharmacology of *T. arjuna* has been mainly due to the tannins present in their barks (Kumar and Prabhakar, 1987).

1A.8 *Ocimum sanctum* (Tulsi Leaves) (Fig. 1A.8)

Plant: *Ocimum sanctum*
Family: Lamiaceae
Genus: *Ocimum*
Part used: leaves

Fig. 1A.8 *Ocimum sanctum.*

Ocimum sanctum (also tulsi or tulasi) is an aromatic plant in the family Lamiaceae, which is native throughout the old world tropics and widespread as a cultivated plant and an escaped weed.[11] It is an erect, much branched subshrub 30–60 cm tall with hairy stems and simple opposite green leaves that are strongly scented. Leaves have petioles and are ovate, up to 5 cm long, usually slightly toothed. Flowers are greenish/purplish in elongate racemes in close whorls. There are two main morphotypes cultivated in India—green-leaved (Sri or Lakshmi tulsi) and purple-leaved (Krishna tulsi).

[10] https://hort.purdue.edu/newcrop/CropFactSheets/terminalia.html
[11] https://en.wikipedia.org/wiki/Ocimum_tenuiflorum

Tulsi is cultivated for religious and medicinal purposes and for its essential oil. It is widely known across South Asia as a medicinal plant and an herbal tea, commonly used in Ayurveda.

Tulsi is mostly used for medicinal purposes and in herbal cosmetics and is widely used in skin preparations due to its antibacterial activity. For centuries, the dried leaves of tulsi have been mixed with stored grains to repel insects. Antimicrobial property of tulsi plant can be used in textile dyeing. If along with natural dyeing some important feature is also added on to the fabric, it will be a real breakthrough. Durable antimicrobial cellulose-containing fabrics have a great deal of demand in a country like ours, where temperate climate conditions especially during rainy season cause immense damage to untreated cotton fabric.

It is very well documented that tulsi leaves have great healing power (Thilagavathi et al., 2005). As it is, naturally dyed materials have good resistance to moth invasion, and to consider, dyeing with tulsi extract will have added advantage because tulsi is known to have antibacterial, antifungal, and antiviral properties. But to have both antifungal treatment and dyeing simultaneously is a major breakthrough for surgical application fabrics. Nonwoven fabrics are needed for surgical gowns, patient's drapes, laboratory coats, coveralls, and other types of protective clothing. During surgical procedures, the personnel may be exposed to sprays of bloods and other fluids that potentially contain blood-borne pathogens. Therefore, surgical gown and patient drape materials should have antimicrobial, not only to reduce infections for patients but also to protect the surgical staff from infection fluids. The color that the tulsi leaves methanolic extract impart to the fabric is a pleasant green color having a faint aroma.

Dyeing along with antimicrobial finish has not been undertaken so far for any of the popular natural dyes; this is the first example of antimicrobial finish along with dyeing fabric or yarn in green color (Tiwari et al., 2000). Although separate application is known in the industry where dyeing is done separately and then antifungal treatment is done on fabric.

1A.9   *Rubia cordifolia* Linn.

Plant: *Rubia cordifolia*
Family: Rubiaceae
Genus: *Rubia*
Part used: roots, stem, and leaves

Rubia cordifolia (local name tamin) produces anthraquinone reddish-orange dyes in roots stem and leaves, which has been used for dyeing textiles since ancient times (Fig. 1A.9). Commercial sonicator dyeing with *Rubia* showed that pretreatment with biomordant, *Eurya acuminata* DC. var. euprista Korth (Theaceae family) (local name Nausankhee (Apatani tribe) and Turku (Nyishi tribe)), in 2% shows very good fastness properties for dyed cotton using dry powder as 10% of the weight of the fabric is optimum. Use of biomordant replaces metal mordants making natural dyeing eco-friendly.

Fig. 1A.9 *Rubia cordifolia.*

The botanical name of manjistha is *R. cordifolia*, and it belongs to family Rubiaceae. It grows throughout India, in hilly districts up to 3500 m height. Generally, this plant is found on the Himalayan side of India. It is a perennial, herbaceous climber. The stems are often long, rough, and grooved, with woody base. The leaves are often in whorls of four. They are 5–10 cm long, variable, cordate-ovate to cordate-lanceolate, rough above, and smooth beneath. The roots are reddish, cylindrical, and flexuous, with a thin red bark and 4–8 cm length.

It is a good source of anthraquinone, reddish-orange dyes, which is obtained from roots, stem, and leaves of the plant. It has been proved to be good and fast dye for fabrics.

Flowers are greenish white or yellowish or red small and sweet scented and found in terminal panicled glabrous cymes. Fruits are globose or slightly lobed and purplish black when ripe.

Roots of *R. cordifolia* are used as natural dye as they contain reddish-orange colorant. The roots contain resinous and extractive matter, gum, sugar, coloring matter, the salt of the pigment being a red crystalline principle "purpurine." The yellow glucoside manjistin and xanthine are also present, besides garancin and orange-red alizarin. Anthraquinone pentacyclic triterpenes, quinines, cyclic hexapeptides, and diethylesters are also reported in literature (Vankar et al., 2008b).

In the Himalayan region, Arunachal Pradesh is recognized as one of the hotspots of biodiversity and the indigenous knowledge system particularly associated with extraction and processing of natural dyes like *R. cordifolia*. These natural dyes are abundantly found there, and from ancient times, some tribes of the state were engaged in natural dyeing. The different tribes mainly the Monpas, Apatanis, Nyishis, and Adis, respectively, of West Kameng, Tawang, Lower Subansiri, and East and West Siang districts of Arunachal Pradesh have been engaged in extraction, processing, and preparation of dyes using barks, leaves, fruits, and roots of the plants from time immemorial.

The natural dyestuffs of plant origins, grown in Arunachal Pradesh, used as indigenous systems can be developed scientifically and can be substituted for the chemical dyes. These indigenous dyes can be produced in large scale and could be prepared commercially and economically. The practice of indigenous systems for preparing dyestuffs and the processes of dyeing has been developed using modern technological methods.

These natural dyes derived from the plants of Arunachal Pradesh are found to be of high quality, and thus, these plants need to be protected for conservation of biodiversity of the flora of northeastern region. People can produce these dyes in large scale, commercially, by establishing processing units and can replace the use of chemical dyes. As Arunachal Pradesh amid its rich diverse flora harbors many dye-yielding plant species in abundance, a study has been carried out to revive and restore the traditional dyeing practices using the traditional biomordant—*E. acuminata* (Nausankhee and Turku)—in place of metal mordant. Innovative methods of dye extraction and mordant study by means of application of modern technology to sharpen the skills of tribal traditional dyers of Arunachal Pradesh, which are hazardous from the environmental point of view, has been focused. Although a lot of work has been done on natural dyeing with *Rubia* (Angelini et al., 1997; Asada et al., 1993; Singh et al., 1993; Bhuyan and Saikia, 2005), the approach toward development of eco-friendly natural dyeing using biomordant and ultrasound energy is comparatively new.

Experience with natural dyeing has given an insight to explore plants in the neighborhood. Finding fiber colors in plants that grow easily and fast has lead in to a new world of fiber colors that give exotic shades. These natural colors have richness and luster that synthetics can never attain. It has become a common misconception that natural dyes only produce beiges and browns and other washed out shades. In reality, vibrant, fast natural colors can be produced, which are comparable with and often surpass the colors of synthetics. Apart from the sources of these dyes, it is perhaps the commitment of those propagating them and the near clinical efficiency with which dye is extracted, produced, and used, which is responsible for the unique nature of natural dyeing and producing stable coloration.

References

Angelini, L.G., Pistelli, L., Belloni, P., Bertoli, A., Panconesi, S., 1997. *Rubia tinctorum* a source of natural dyes: agronomic evaluation, quantitative analysis of alizarin and industrial assays. Ind. Crop. Prod. 6, 303–311.

Asada, H., Nakamura, R., Torimoto, N., Takaoka, A., 1993. Akane (*Rubia cordifolia* L. var. munjist Mig) as a teaching material. Kagaku to Kyoiku 4, 339–342.

Bhuyan, R., Saikia, C., 2005. Isolation of colour components from native dye-bearing plants in northeastern India. Bioresour. Technol. 96, 363–372.

Ghorpade, B., Tiwari, V., Vankar, P.S., 2000. Ultrasound energised dyeing of cotton fabric with canna flower extracts using eco friendly mordants. Asian Text. J. 3 (March), 68–69.

Kale, S., Naik, S., Deodhar, S., 2006. Utilization of *Cosmos sulphureus* Cav. flower dye on wool using mordant combinations. Nat. Prod. Rad. 5, 19–24.

Kumar, D.S., Prabhakar, Y.S., 1987. On the ethnomedical significance of the arjun tree, *Terminalia arjuna* (Roxb.) Wight & Arnot. J. Ethnopharmacol. 20, 173–190.

Mahanta, D., Tiwari, S.C., 2005. Natural dye yielding plants and indigenous knowledge on dye preparation in Arunachal Pradesh, northeast India. Curr. Sci. 88, 1474.

Mohanty, B.C., Chandramouli, K.V., Naik, H.D., 1987. Natural Dyeing Processes of India. Calico Museum of Textiles, Sarabhai Foundation, Ahmedabad.

Singh, S., Jahan, S., Gupta, K., 1993. Optimization of procedure for dyeing of silk with natural dye madder roots (*Rubia cordifolia*). Colourage 40, 33–36.

Srivastava, J., Seth, R., Shanker, R., 2008. PS Vankar solubilisation of red pigments from Canna Indica flower in different media and cotton fabric dyeing. International Dyer 193 (1), 31–36.

Thilagavathi, G., Rajendrakumar, K., Rajendran, R., 2005. Development of ecofriendly antimicrobial textile finishes using herbs. Indian J. Fibre Text. Res. 30, 431–436.

Tiwari, V., Ghorpade, B., Mishra, A., Vankar, P.S., 2000. Ultrasound dyeing with *Ocimum sanctum* (Tulsi leaves) with ecofriendly mordants. New Cloth Market 1, 23–24.

Tiwari, V., Shanka, R., Vankar, P.S., 2001. Ultrasonic dyeing cotton fabric with Babool bark (*Acacia arabica*) for preparation of eco-friendly textile. Chem. World 31 (March), 30–32.

Usher, G., 1974. A Dictionary of Plants Used by Man. Constable and Company Ltd, London.

Vankar, P., 2002. Commercial viability of natural dyes: Henna, harda, catechu and babool for textile dyeing. Nat. Prod. Rad. 1, 15–18.

Vankar, P.S., Kushwaha, A., 2011. Salvia splendens, a source of natural dye for Cotton and Silk fabric dyeing. Asian Dyers 6, 29–32.

Vankar, P.S., Shanker, R., 2010. Natural dyeing of Silk and Cotton by Rhododendron flower extract. Int. Dyers 7, 37–40.

Vankar, P.S., Shanker, R., Dixit, S., Mahanta, D., Tiwari, S., 2008a. Sonicator dyeing of modified cotton, wool and silk with *Mahonia napaulensis* DC. and identification of the colorant in Mahonia. Ind. Crop. Prod. 27, 371–379.

Vankar, P.S., Shanker, R., Mahanta, D., Tiwari, S., 2008b. Ecofriendly sonicator dyeing of cotton with *Rubia cordifolia* Linn. using biomordant. Dyes Pigments 76, 207–212.

Zhu, M., Sibiao, L., 1991. The effect of metal ions on mulberry and Rhododendron pigments. Shipin Yu Fajiao Gongye 6, 82–83.

Description of the newer natural dye sources suitable for silk fabrics

P.S. Vankar
FEAT (Facility for Ecological and Analytical Testing), Kanpur Kalyanpur, India

Introduction

Demand for cheap natural dyes with good fastness properties in all dyeing industries globally has increased. Worldwide, growing consciousness about organic value of eco-friendly products has generated renewed interest of consumers toward use of textiles (preferably natural fiber product as silk) dyed with eco-friendly natural dyes.

Studies on silk

Pure munga khadi silk and plain and white pure silk, which is generally used for dyeing, were selected. Standard brand of pure silk was procured from the market. Various natural dyes were used to dye pure silk to get many bright colors.

1B.1   Black carrot/*Dacus carota*

Plant: *Daucus carota*
Family: Umbelliferae
Genus: *Daucus*
Part used: root

Natural dyes from plants are getting increasing attention as an alternate source for synthetic dyes in the food (Wissgott and Bortlik, 1996), textile (Deo and Desai, 1999), and pharmaceutical industry (Vankar and Srivastava, 2008), and they increase their added value if they possess positive effects on health (Vankar and Srivastava, 2010) and some added qualities like antifungal properties (Bajpai and Vankar, 2007).

Black carrot has been investigated as a dye source. It has been used for conventional dyeing along with metal mordants, and attempts have been made to develop methods to optimize its dyeing characteristics as a natural dye source. It is a cheap

Natural Dyes for Textiles. http://dx.doi.org/10.1016/B978-0-08-101274-1.00007-0

Fig. 1B.1 Black Carrot root.

and abundantly available dye source. On a different research line, this work has been focused on the development of new dye source and rapid methods for the effective dyeing with Black carrot or *D. carota* (Fig. 1B.1).

Black carrot (*D. carota* ssp. *sativus* var. *atrorubens*) grows mostly in southern Europe and Asia, principally in Turkey and India. The pigments that give black carrots their characteristic color are anthocyanins. Anthocyanins from black carrot are more stable over a wider pH range than anthocyanins from other fruit or vegetable sources, making them ideal for use in food products website.[1] This fact was exemplified by another study (Jackman et al., 1987) in which the black carrot extract and its isolated anthocyanins at three pH values showed that cyanidin 3-xylosyl (sinapoylglucosyl) galactoside was found to exhibit a lower visual detection threshold and a higher pH stability than cyanidin 3-xylosyl (feruloylglucosyl) galactoside and cyanidin 3-xylosyl (coumaroylglucosyl) galactoside.

The silk dyeing with black carrot was attempted for silk fabric. Black carrot or *D. carota ssp. sativus var. atrorubens* has not been used for textile dyeing so far. This newly claimed natural dye has attributed very good shades on silk. The colorant responsible for beautiful pastel shades of green was primarily anthocyanins. Anthocyanins from black carrot are more stable over a wider pH range than anthocyanins from other fruit or vegetable sources, making them ideal for use. Premordanted with metal salts, conventionally dyed silk has given impressive CIE $L^*a^*b^*$ values. The K/S of silk increased in the order of the dyeing using the mordants—$SnCl_2 >$ FeSO4 $>$ SnCl4 $>$ Alum $>$ $CuSO_4 >$ $K_2Cr_2O_7$ mordants. Fastness properties have also been good with silk swatches. Black carrot grows mainly in Asian region and can be a potential dyeing source for natural dyeing lovers (Shukla and Vankar, 2013).

[1] http://www.carrotmuseum.co.uk/blackcarrot.html.

1B.2   *Hibiscus-rosa sinensis*

Plant: *Hibiscus rosa-sinensis*
Family: Malvaceae
Genus: *Hibiscus*
Part used: flower (red)

Fig. 1B.2 Red Hibiscus flower.

Hibiscus rosa-sinensis, also known by the common name red hibiscus, is a large shrub or small tree that grows up to 4.7 m tall (Fig. 1B.2). The genus *Hibiscus* contains approximately 200 species distributed throughout tropical and subtropical regions.[2]

This plant has a variable stature and may be upright or broad and spreading. The leaves are arranged alternately on the branches and are ovate in shape (wider at the base than at the tip) and grow from 5 to 15 cm long. The red flowers are very large and can be up to 15 cm long. Like all *Hibiscus* flowers, the stalks of the stamens (the pollen producing structures) and the style are fused into a long column that is exerted from the center of the widely spreading petals. In cultivated varieties, the petals may be single or double and smooth or scalloped. Many anthers (in which the pollen is produced) can be seen partway up the column, and five round stigma lobes (onto which the pollen lands in order for fertilization to occur) are visible at the tip of the column.

Anthocyanins are natural colorants that have extensive range of colors and occur widely in nature. Anthocyanins are the most important pigments ranging from orange, pink, red, and violet to blue in the flowers and fruits of the vascular plants. They are harmless and water-soluble, which makes them interesting for their use as natural water-soluble colorants. Despite the great potential of applications that anthocyanins represent for food, pharmaceutical,

---

[2] https://en.wikipedia.org/wiki/Hibiscus_rosa-sinensis.

and cosmetic industries, their use has been limited because of their relative instability and low extraction percentages (Castaneda-Ovando et al., 2009). Their use in textile is negligible as they lack affinity for the fiber and cannot sustain washing. Nevertheless, anthocyanins are good food colorants, because in those applications color fastness properties do not play such an important role as for the textile applications. Currently, most investigators are engaged in solving the problems that are associated with isolation and stability of anthocyanins, their purification, their identification, and their end uses. Dyeing wool with *H. rosa-sinensis* flower was also attempted (Shanker and Vankar, 2007). Anthocyanin from hibiscus flower was extracted, and silk was dyed using different metal mordants.

Red hibiscus flower is used for dyeing (Vankar and Shukla, 2011). The anthocyanin present in hibiscus was extracted, and primarily silk was dyed with the extract. The extraction of anthocyanins using ethanol acidified with citric acid (0.01%) instead of hydrochloric acid was reported (Castaneda-Ovando et al., 2009; Main et al., 1978). Citric acid is less corrosive than hydrochloric acid, chelates metals, maintains a low pH, and may have a protective effect during processing (Metivier et al., 1980).

1B.3   *Delonix regia*

Plant: *Delonix regia*
Family: Fabaceae
Genus: *Delonix*
Part used: flower

Fig. 1B.3 Delonix flower.

Delonix belongs to the family Leguminaceae (Fig. 1B.3). It is a flamboyant, royal poinciana, flame tree that grows rapidly up to 20 m high with bipinnate leaves 15–20 cm long and leaflets 0.5–1 cm long. It grows in temperate condition in abundance in the northern parts of India. The red flowers are 3–4 cm in diameter and grow on terminal branches.[3] *Delonix* (Gulmohar) is a flamboyant tree in flower—some say the world's most colorful tree. For several weeks in spring and summer, it is covered with exuberant clusters of flame-red flowers, 4–5 in. across. Bunches of scarlet red flowers

---

[3] https://en.wikipedia.org/wiki/Delonix_regia.

appear in April to July when the tree becomes leafless. Flowers are large brilliant red-orange, occurring in numerous, huge terminal clusters at the ends of branches; each individual flower has five large, wide-spreading petals (each 1 1/2–2 in. long), one petal streaked with white and yellow, and flower appears in early summer and continues for several months in Indian temperate climate. Thus, *Delonix* flower, though seasonal, is abundantly available for dyeing purpose. The flame-colored flowers are formed in dense clusters and bloom seasonally, usually in midsummer. The individual flowers are between 8 and 15 cm in diameter and have four spoon-shaped spreading scarlet or orange-red petals (4–7 cm long) and one upright slightly larger petal (the standard) that is marked with yellow and white. The dark-brown pods are flat and woody, up to 70 cm long and 7 cm wide. They contain between 18 and 45 yellowish to dark-brown seeds that are about 2 cm long. Scientists have used different parts such as petal, calyx, and whole flowers of *D. regia* for cotton and silk yarn dyeing using metal mordant (Purohit et al., 2007). It is for the first time *Delonix* flower extract has been used in conjunction with a biomordant and enzymes for silk dyeing.

The red flowers of *D. regia* have been evaluated for natural dyeing of silk using biomordant and enzymes for the first time, with a deliberate attempt to avoid metal mordanting in silk dyeing (Vankar and Shanker, 2009). This would make textile dyeing more eco-friendly. The study was designed to evaluate the potential of this natural dye source for its dye content and to replace metal mordanting by the use of enzyme or biomordants.

1B.4 *Plumeria rubra* (pink)

Plant: *Plumeria rubra*
Family: Apocynaceae
Genus: *Plumeria*
Part used: flower

Fig. 1B.4 Plumeria flower.

Plumeria (pink champa) is native to warm tropical areas of India (Fig. 1B.4). Plumeria can grow to be large shrubs or even small trees. In tropical regions, plumeria may reach a height of 30–40 ft and half as wide.[4] They have widely spaced thick succulent branches, round or pointed, long leather, fleshy leaves in clusters near the branch tips. Sensitive to cold, leaves tend to fall in early winter since they are deciduous. During the early summer, the very fragrant clusters of showy, waxy flowers fill the shrub. There is absolutely nothing like the sweet fragrance of plumeria in flower, with fragrances of jasmine, citrus, spices, gardenia, and other indescribable scents. These flowers are known for their durability, fragrances, and colors of whites, yellows, pinks, reds, and multiple pastels. Flowering can last up to 3 months at a time producing new blooms every day.

Plumeria rubra grows as a spreading shrub or small tree to a height of 2–8 m (5–25 ft) and similar width. It has a thick succulent trunk and sausage-like blunt branches covered with a thin gray bark. The branches are somewhat fragile and brittle and, when broken, ooze a white latex that can be irritating to the skin and mucous membranes.[5] The large green leaves can reach 30–50 cm (12–20 in.) long and are arranged alternately and clustered at the end of the branches. They are deciduous, falling in the cooler months of the year. The flowers are terminal, appearing at the ends of branches over the summer. Often profuse and very prominent, they are strongly fragrant and have five petals. The colors range from the common pink to white with shades of yellow in the center of the flower. Initially tubular before opening out, the flowers are 5–7.5 cm (2–3 in) in diameter.

The flowers are clustered at the branch tips. The individual flowers are tubular, 2 in. (5 cm) across, and have five broadly to narrowly oval lobes with yellow at their base. The flower stalks, the flower buds, and the outside of the petals are reddish or tinged with red. The flowers emerge before the leaves in the springtime.

Once picked, a bloom can last for several days without wilting if kept in water. Plumeria dark-pink variety was chosen for the dyeing study. The shrub grows best in full sun in deep, rich, well-drained soils. A reddish pink dye is obtained from the petals. The study reports that the flowers can be used for coloring purposes in both fresh and dry forms.

Coloring pigment from plumeria or pink champa flower has been extracted and used for dyeing silk fabric. It is observed that the dyeing with dark-pink flowers gives fair-to-good fastness properties in sonicator dyeing method in just 1 h and shows good dye uptake as compared with conventional dyeing (Vankar and Shankar, 2007). The pigments found in plumeria are rutin and quercitrin. The shade ranges from green to purplish gray in the presence of 2%–4% of metal mordants.

1B.5   *Combretum indicum*

Plant: *Combretum indicum*
Family: Combretaceae
Genus: *Combretum*
Part used: flower

---

[4] https://en.wikipedia.org/wiki/Plumeria_rubra.
[5] https://www.flowersofindia.net/catalog/slides/Red%20Frangipani.html.

Fig. 1B.5 *Combretum indica* flower.

Combretum indicum is also known as the Rangoon creeper/Madhumalti (Fig. 1B.5). It is a vine with dark-pink flower clusters and is found in Asia. It is found in many other parts of the world, but it is a very popular plant in India.[6] The creeper is a ligneous vine that can reach up to 8 m. The flowers are fragrant and tubular, and their color varies from white, to light pink, to dark-pink varieties all growing in the same bunch. *Combretum indicum* belongs to the family Combretaceae. It is a large woody plant. It is indigenous to tropical Africa and Indo-Malaysian region. It has been probably introduced into India, as it is not found growing wild anywhere. It is a hardy creeper commonly planted in gardens for the brightly colored showy flowers. The plant blooms profusely during two seasons March–May and September–November though it flowers throughout the year. Flowers appear in constant succession in drooping clusters; they open in the evening as white flowers, gradually assuming a pink tinge by morning and deepening to deep red by late afternoon. The flowers have a characteristic mild floral sweet astringent fragrance pervading into the surroundings.

The conventional dyeing of silk fabrics with the aqueous extract of dark-pink variety flowers of *C. indicum* by using metal mordants—alum, ferrous sulfate, and potassium dichromate. CIE L*a*b* values, K/S values, and fastness properties of the dyed fabrics were ascertained. The dye showed promising results and acceptability for commercial dyeing. This is a report on the use of *C. indicum* as natural colorant (Vankar and Srivastava, 2011).

---

[6] https://en.wikipedia.org/wiki/Combretum_indicum.

1B.6   *Ixora coccinea*

Plant: *Ixora coccinea*
Family: Rubiaceae
Genus: *Ixora*
Part used: red flowers

Fig. 1B.6 *Ixora coccinea.*

Ixora coccinea is a traditional medicinal plant, cultivated throughout India (Fig. 1B.6). It is extensively used in Ayurvedic medicine.[7] A variety of chemical compounds have been isolated from this plant. The various medicinal uses of this plant, its chemical, pharmacognosy, and pharmacology are known in the literature. Ixora grows as shrub that is 4–8 ft high and is also known by common names such as flame of the woods, jungle flame, crimson king, or jungle geranium, which is a bushy, rounded shrub that has long been a popular hedging plant in some subtropical regions. Ixoras are compact plants that bloom primarily in summer and intermittently the rest of the year with proper care. Ixoras freely produce loose, corymb-like cymes, 2–5 in. across, of red, orange, pink, or yellow flowers. Available in a number of cultivars, this plant is a member of the Rubiaceae family. The blooms are in a bunch of different colors such as pink, white, peach, and red. The single flowers have four petals. Plants will grow and flower in shade, but most ixoras do best in full sun, acid soil, and well-drained moist organic mix. A reddish dye is obtained from the petals. The study reports that the flowers can be used for coloring purposes, in both fresh and dry forms.

Coloring pigment from Ixora flower has been extracted and used for dyeing silk fabrics. It is observed that dyeing with ixora gives fair-to-good fastness properties in sonicator in 1 h and shows good dye uptake as compared with conventional dyeing (Vankar and Shanker, 2006). The pigment is found to contain chrysin 5-O-β-D-xylopyranoside and flavone, 4′,5,7-trihydroxyflavone-4′-β-D-glucopyranoside. The shade ranges from green to purple in the presence of 2%–4% of mordants.

---

[7] https://en.wikipedia.org/wiki/Ixora_coccinea.

1B.7   *Bischofia javanica*

Plant: *Bischofia javanica*
Family: Euphorbiaceae
Genus: *Bischofia*
Part used: leaves

Fig. 1B.7 *Bischofia javanica Bl.*

Bischofia javanica Bl. (local name Maub) belongs to family Euphorbiaceae and produces natural dye that has been used for dyeing textiles by some tribes of Arunachal Pradesh. *Bischofia javanica* is an evergreen woody tree with a maximum height of 40 m and diameter 2.3 m.[8] It occurs in humid valley forest of Arunachal Pradesh. Stem and leaves bear natural dye (Fig. 1B.7). The leaves are trifoliate with petiole 8–20 cm in length. The different tribes of Arunachal Pradesh have been engaged in extraction, processing, and preparation of dyes using barks, leaves, fruits, and roots of the plants from time immemorial.

Bischofia javanica leaves have been used as natural dyestuff used as indigenous systems, which has been developed scientifically and can be substituted for the chemical dyes. This indigenous dye can be produced in large scale and could be prepared commercially and economically. The practice of indigenous systems for preparing dyestuffs and the processes of dyeing has been developed using modern technological methods.

If the indigenous dyes derived from the plants of Arunachal Pradesh are found to be of high quality, these plants can be protected for conservation of biodiversity of the flora of northeastern region. People can produce these dyes in large scale, commercially,

[8] https://en.wikipedia.org/wiki/Bischofia_javanic.

by opening factories and can compete with chemical dyes that are harmful in environmental point of view. As Arunachal Pradesh with its diverse flora is a resource base of
dye-yielding plant species, we carried out a study to revive and restore the traditional
dyeing by using innovative technology. This work was designed with an aim to focus
on the innovative methods of dye extraction as well as the use of least amount of mordant and betterment of fastness properties in dyed fabric.

Innovative sonicator dyeing with *Bischofia* shows good result. The color extracted
from the leave was dark brown (Vankar et al., 2007). Pretreatment with 2% metal
mordant and using 10% of plant extract for the weight of the fabric was found to be
optimum showing very good fastness properties for both cotton and silk dyed fabrics.

References

Bajpai, D., Vankar, P.S., 2007. Antifungal textile dyeing with *Mahonia napaulensis* DC leaves
    extract based on its antifungal activity. Fibers Polym. 8, 487–494.
Castaneda-Ovando, A., De lourdes Pacheco-Hernández, M., Páez-Hernández, M.E., Rodríguez,
    J.A., Galán-Vidal, C.A., 2009. Chemical studies of anthocyanins: a review. Food Chem.
    113, 859–871.
Deo, H., Desai, B., 1999. Dyeing of cotton and jute with tea as a natural dye. Color. Technol.
    115, 224–227.
Jackman, R.L., Yadav, R.Y., Tung, M.A., Speers, R., 1987. Anthocyanins as food colorants—a
    review. J. Food Biochem. 11, 201–247.
Main, J., Clydesdale, F., Francis, F., 1978. Spray drying anthocyanin concentrates for use as
    food colorants. J. Food Sci. 43, 1693–1694.
Metivier, R., Francis, F., Clydesdale, F., 1980. Solvent extraction of anthocyanins from wine
    pomace. J. Food Sci. 45, 1099–1100.
Purohit, A., Mallick, S., Nayak, A., Das, N., Nanda, B., Sahoo, S., 2007. Developing multiple
    natural dyes from flower parts of Gulmohur. Curr. Sci. 92, 1681–1682.
Shanker, R., Vankar, P.S., 2007. Dyeing wool yarn with *Hibiscus rosa sinensis* (Gurhhal).
    Colourage 54, 66–69.
Shukla, D., Vankar, P.S., 2013. Natural dyeing with black carrot: new source for newer shades
    on silk. J. Nat. Fibers 10, 207–218.
Vankar, P.S., Shanker, R., 2007. Dyeing silk and wool with Plumeria (pink) flower. Asian Text.
    J., 104–107.
Vankar, P., Shanker, R., 2006. Sonicator dyeing of cotton and silk fabric by *Ixora coccinea*.
    Asian Text. J. 2, 77–80.
Vankar, P.S., Shanker, R., 2009. Potential of *Delonix regia* as new crop for natural dyes for silk
    dyeing. Color. Technol. 125, 155–160.
Vankar, P.S., Shukla, D., 2011. Natural dyeing with anthocyanins from *Hibiscus rosa sinensis*
    flowers. J. Appl. Polym. Sci. 122, 3361–3368.
Vankar, P.S., Srivastava, J., 2008. Comparative study of total phenol, flavonoid contents and
    antioxidant activity in *Canna indica* and *Hibiscus rosa sinensis*: prospective natural food
    dyes. Int. J. Food Eng. 4.
Vankar, P.S., Srivastava, J., 2010. Ultrasound-assisted extraction in different solvents for phyto-
    chemical study of *Canna indica*. Int. J. Food Eng. 6, 1–11.

Vankar, P.S., Srivastava, J., 2011. Natural dyeing with dark pink flowers of *Quisqualis indica* (*Combretum indicum*). Asian Dyers 58, 61–63.

Vankar, P.S., Shanker, R., Mahanta, D., Tiwari, S.C., 2007. Characterization of the colorants from leaves of *Bischofia javanica* Bl. and Sonicator dyeing of cotton and silk with the extract. Int. Dyer 192, 31–33.

Wissgott, U., Bortlik, K., 1996. Prospects for new natural food colorants. Trends Food Sci. Technol. 7, 298–302.

Description of the newer natural dye sources suitable for wool yarn

P.S. Vankar
FEAT (Facility for Ecological and Analytical Testing), Kanpur Kalyanpur, India

1C.1 *Celosia cristata*

Plant: *Celosia cristata*
Family: Amaranthaceae
Genus: *Celosia*
Part used: magenta flowers

Celosia cristata (murgkesh), mainly betalains, is an unexplored class of natural dye for wool yarn dyeing. *C. cristata* belongs to the family Amaranthaceae; *Celosia* blooms for longer time. Recently, there has been renewal of interest in this exotic group of plants (Shanker et al., 2004; Shanker and Vankar, 2005c; Shanker and Vankar, 2005b) for several reasons. Every bright-colored flora is being considered for natural dyeing studies. The aim to use this plant source has been to identify newer sources of natural dyes and based on the chemical structure of the colorant. Innovative pretreatment and mordants for better dye uptake by the wool yarn have been used. Deep-red-colored celosia flowers grow in full sun in a rich amended soil for full bloom. The blooming period is August–November and lasts long, as the flowers have velvety petals that can with stand dry weather.

Celosia is an herb, grows widely or cultivated, and is distributed throughout the tropical and temperate regions of India. Four species are popularly grown in India. One of the species of *C. cristata* locally known as lal murgh-kesh (Fig. 1C.1), a showy plant often cultivated for ornamental purposes in gardens and as an escape in the plains and up to a height of 5000 ft on the Himalayas. The plant is often eaten as a pot herb. The plant yields betacyanin, a nitrogen-containing anthocyanin. The flowers are astringent and are used in diarrhea. The seeds are demulcent and useful in cough and dysentery. Color is an important constituent of food and pharmaceuticals as every food is associated with a certain type of color. Color is necessary to give food attracting and appetizing appearance. Red-violet pigments called betacyanins like betanidin, the aglycone of red-violet pigment betanin isolated originally from *Beta vulgaris*. A violet-red dye, which is an extract of the flowers of *Celosia cristata*, can be used in food stuffs like other betacyanins, as this plant is eaten as pot herb. Betacyanin is the

Natural Dyes for Textiles. http://dx.doi.org/10.1016/B978-0-08-101274-1.00008-2

Fig. 1C.1 Celosia flower.

major component in *C. cristata* flowers responsible for violet-red color.[1] The known process for the extraction of dye showed the presence of celosianin and isocelosianin (both are C-15 diastereomers), and by alkaline hydrolysis of the celosianin, isocelosianin mixture gives a deacylated product that was identified as a mixture of amarantin and isoamarantin (Minale et al., 1966).

This method provided an improved process for the extraction of betacyanin dye from the *C. cristata* flowers to obtain betacyanin dye with the total dry mass of the dye being in the range of 15%–22% on dry weight basis. This source has been used for dyeing wool yarn using metallic salts such as alum, stannic chloride, stannous chloride, and ferrous sulfate as mordants. Ethylene diamine, sodium hydroxide, and morpholine were used as pretreatments for the wool (Shanker and Vankar, 2005a).

1C.2   *Nerium oleander*

Plant: *Nerium oleander*
Family: Apocynaceae

[1] http://www.allindianpatents.com/patents/194605

Fig. 1C.2 Nerium flower.

Genus: N*erium*
Part used: dark pink flowers

Nerium oleander is an evergreen shrub reaching 4 m in height. Leaves are 10–22 cm long, narrow, untoothed and short-stalked, dark or gray-green in color (Fig. 1C.2). Some cultivars have leaves variegated with white or yellow.[2] All leaves that have a prominent midrib are "leathery" in texture and usually arise in groups of three from the stem. The plant produces terminal flower heads, usually pink or white; however, 400 cultivars have been bred, and these display a wide variety of different flower colors: deep to pale pink, lilac, carmine, purple, salmon, apricot, copper, orange, and white. Each flower is about 5 cm in diameter and five-petalled. The throat of each flower is fringed with long petallike projections. Occasionally, double flowers are encountered among cultivars. The fruit consists of a long narrow capsule 10–12 cm long and 6–8 mm in diameter; they open to disperse fluffy seeds. Fruiting is uncommon in cultivated plants. The plant exudes a thick white sap when a twig or branch is broken or cut. Where the species grows in the wild, it occurs along watercourses, gravely places, and damp ravines. It is widely cultivated particularly in warm temperate and subtropical regions where it grows outdoors in parks, gardens, and along road sides.

N. oleander flower has been used in the ultrasonic dyeing of wool. It was found that *Nerium* flowers give color in hot water very quickly. Moreover, it was observed that dyeing with *Nerium* provided fair-to-good fastness properties in sonicator dyeing (Vankar and Shanker, 2008).

---

[2] https://en.wikipedia.org/wiki/Nerium

References

Minale, L., Piattelli, M., De Stefano, S., Nicolaus, R.A., 1966. Pigments of centrospermae—VI. Phytochemistry 5, 1037–1052.

Shanker, R., Vankar, P.S., 2005a. Dyeing with Celosia cristata flower on modified pretreated wool. Colourage 52, 53–56.

Shanker, R., Vankar, P.S., 2005b. Dyeing wool with Gomphrena globosa flower. Colourage 52, 35–38.

Shanker, R., Vankar, P.S., 2005c. Ultrasonic energised dyeing of wool with Mirabilis jalpa flowers. Colourage 52, 57–61.

Shanker, R., Vankar, S., Vankar, P.S., 2004. Ultrasound energised dyeing of wool with portulaca flower extracts using metal mordants. Colourage 51, 41–46.

Vankar, P.S., Shanker, R., 2008. Ultrasonic dyeing of cotton and silk with *Nerium oleander* flower. Colourage 55, 90–94.

Special case study: Dyeing of cotton, silk, and wool with natural dyes

P.S. Vankar
FEAT (Facility for Ecological and Analytical Testing), Kanpur Kalyanpur, India

1D.1 *Alcea rosea*

Plant: *Alcea rosea*
Genus: Alcea
Family: Malvaceae
Part used: Flower

Fig. 1D.1 *Alcea rosea.*

Natural Dyes for Textiles. http://dx.doi.org/10.1016/B978-0-08-101274-1.00013-6

Hollyhock (*Alcea rosea*) grows 4–8 ft high; they grow best in full sun in deep, rich, and well drained soils (Fig. 1D.1). Some varieties act as reseeding biennials.[1] The blooms are dramatic spires of rosette, single, or double flowers in a scope of colors. The single flowers have five petals. Many colors are available from pastel pinks to near black. A reddish dye is obtained from the petals (Ferenczi et al., 1981; Kasumov, 1984). The study reports that the flowers can be used for food coloring purposes both in fresh and dry forms. The red anthocyanin constituent of the flowers is known to be used as litmus (Matula and Macek, 1936).

The flowers of all colors will make a dark pink dye on fibers mordanted with alum and crème of tartar. The black variety of flowers produces purplish shade. Changing the pH of the dye bath provides even greater color variations, including blue, green, and brown. Fresh leaves of the plants have also been used to produce green dyes.

Flowers from plant source were crushed and dissolved in distilled water and allowed to boil in a beaker kept over a water bath for quick extraction for 3 h. All the color was extracted from flowers by the end of 3 h. The solution was filtered for immediate use. The flowers were also dried in trays, in thin layers, in a current of warm air immediately after picking. When dry, they are a deep, purplish-black. These flowers could be used as and when required. The colorant showed one major peak, max at 296.33 nm in the UV region (flavonoids) and at 547.96 nm.

Coloring pigment from Hollyhock (*Alcea rosea*) flower has been extracted and used for dyeing wool yarn, silk, and cotton fabrics. It is observed that the dyeing with hollyhock gives fair to good fastness properties in sonicator in 1 h and shows good dye uptake as compared with conventional dyeing. The pigment is found to contain cyanidin-3-glucoside, delphinidin-3-glucoside, and malvidin-3, 5-diglucoside. The shade ranges from green to brown in the presence of 2%–4% of mordants (Vankar and Shanker, 2006b).

1D.2   *Hibiscus mutabilis* (Gulzuba)

Plant: *Hibiscus mutabilis*
Family: Malvaceae
Genus: Hibiscus
Part used: Flower

The Gulzuba/Cotton rose/*Hibiscus mutabilis* is a large shrub or small multistemmed tree that grows to 15 ft (4.6 m) high with about a 10 ft (3 m) spread (Fig. 1D.2). Neither a true hibiscus nor a rose (it's in Malvaceae, the hibiscus family). *Hibiscus mutabilis* is downright conspicuous when in full bloom starting in late summer and on into fall. The flowers open pure white and change color over a 3-day period until they are deep pink and then as they die assume a dark "blue-pink" hue.[2] The most notable characteristic of this flowering shrub is that flowers of three distinct colors appear on

---

[1] http://www.desert-tropicals.com/Plants/Malvaceae/Alcea_rosea.html.
[2] https://en.wikipedia.org/wiki/Hibiscus_mutabilis.

Fig. 1D.2 *Hibiscus mutabilis.*

the bush simultaneously as the blooms color cycle independent of one another. Single and double flowered varieties are available, both having quite large blossoms that are 3–5 in. (8–13 cm) across. After flowering a round, hairy capsule forms which dries and releases fuzzy seeds, a trait that inspired one of the plants common names, rose cotton as the buds resemble the boll of that famous member of the hibiscus family. The large leaves are 5–7 in. (13–18 cm), bright green, hairy on the undersides and deeply lobed. They impart a coarse texture that gives the plant a distinctive eye-catching appeal. There is always great demand in garden centers for the cotton rose when it is in full bloom, for it is one of the most imposing and unusual of flowering trees. Little to no care is required as this shrub truly takes care of itself and is adaptable to most locations and soil conditions. Sun or light shifting shade. This shrub thrives on regular watering but this is optional as it is very drought tolerant and propagate by cuttings, gulzuba is easy to root.

1D.2.1   Color changes in the petals

The petals of this plant change 1–3 from ivory white in the morning to light rose at noon and pink-red in the evening. Examination of petals at the three periods showed that the flavonol glycosides in all three were isoquercitrin, hyperoside, rutin, quercitin 4′-glucoside, and quercimeritrin, and the free aglycon was quercitin. There was no anthocyanin in the morning batch but in the other two there were cyanidin 3, 5-diglucoside, and cyanidin 3-rutinoside-5-glucoside, with cyanidin as the free aglycon. The content of total anthocyanin in the evening was threefold greater than that at noon. Since during the day there was no decrease in flavonols, the anthocyanins were evidently synthesized independently. When opening in the morning, flowers of *H. mutabilis* appear white or ivory. The flower color changes to red by late afternoon due to the accumulation of the anthocyanin cyanidin-3-sambubioside. At the onset, and during the rapid phase of pigment accumulation, phenylalanine ammonia-lyase (PAL) activity in the petals increases rapidly to seven times its initial level and then decreases while the flower senesces. Anthocyanin synthesis depends, therefore, on the de novo production of cinnamic acid. Two kinds of antho-cyanins in the red petals of *H. mutabilis* were identified as cyanidin 3-xylosylglucoside (ilicicyanin) and cyanidin 3-monoglucoside (chrysanthemin), which were present in the ratio 8:2. The hydrolyzed petal extracts from both white and red flowers contained quercetin and kaempferol in the ratio 7:3. The white petals produced a considerable amount of anthocyanins, when they were detached and floated on water in the light and (or) dark. Metabolic inhibitors inhibited markedly the pigment production in the detached petals, and sucrose promoted it to some extent (Subramanian and Nair, 1970).

Hibiscus mutabilis (Gulzuba)/Cotton rose/ belonging to family Malvaceae produces natural dye which has been used for dyeing textiles. Aqueous extract of Gulzuba flowers yields shades with good fastness properties. The dye has good scope in the commercial dyeing of cotton, silk for garment industry, and wool yarn for carpet industry. In the present study dyeing with gulzuba has shown to give good dyeing results. Pretreatment with 2%–4% metal mordants and keeping M:L ratio as 1:40 for the weight of the fabric to plant extract is optimum showing very good fastness properties for cotton, silk, and wool dyed fabrics (Shanker and Vankar, 2007).

1D.3   *Cayratia carnosa Gagn.* or *Vitis trifolia*

Plant: *Cayratia carnosa*
Family: Vitaceae
Genus: Cayratia
Part used: berries

Fig. 1D.3 *Cayratia carnosa.*

Recently there has been renewal of interest in this exotic group of plants for several reasons (Fig. 1D.3). Every bright colored flora is being considered for natural dyeing studies. Our aim has been to identify newer sources of natural dyes and based on the chemical structure of the colorant we have tried to use innovative pretreatment and mordants for better dye uptake by both the fiber and the fabric. The wildly grown plant called *Cayratia carnosa*, is of family Vitaceae whose ripened fruits were used for dyeing cotton, silk, and wool (Vankar and Shanker, 2006a,b).

The plant is found throughout Northern India in thickets at low altitudes. This vine climbs by means of tendrils, which are found opposite the leaves. The leaves are trifoliolate with petioles 2–3 cm long. The leaflets are ovate to oblong-ovate, 2–8 cm long, 1.5–5 cm wide, pointed at the tip, and coarsely toothed at the margins. The flowers are small greenish white and borne on axillary solitary cymes. The fruit is fleshy, juicy, dark purple or black, and subglobose and about 1 cm in diameter.[3] Deep blue/purple colored berries of Cayratia were collected for dyeing purpose.

1D.4 *Tegetes erecta*

Plant: *Tegetes erecta*
Family: Asteraceae
Genus: Tegetes
Part used: Flower

Fig. 1D.4 *Tegetes* flower.

Tegetes erecta (Marigold) belongs to the family Asteraceae (Fig. 1D.4). It is a small shrub and bears yellowish orange flowers in abundance during the flowering season,

---

[3] https://en.wikipedia.org/wiki/Cayratia_trifolia).

which lasts for more than 6–8 months. It occurs in humid climate in different parts of India and even in Srilanka. *Tegetes* pigments mainly consists of carotenoids and flavonoids, these have been used mainly as natural food colorants and feed (Vasudevan et al., 1997). *Tegetes erecta* L. flower pigments have been extracted and used as a natural food additive to color egg yolks orange and poultry skin yellow (Bosma et al., 2000). Five cultivars were examined for lutein production on commercial scale. "E-1236" (a variety of *Tegetes*) was found to produce the largest quantity of lutein, 22.0 kg ha^{-1}, and "Orange Lady" (another hybrid variety) produced 21.3 kg ha^{-1}.

The flowers are offered to the deities in temples and are thus available in huge quantities as temple waste (Vankar, 2009). The total waste generated in Kanpur city alone has been estimated to be about 20 tonnes per day. Most of these flowers are either dumped by the side of river Ganga or allowed to naturally decay and used as compost. This waste material was collected and utilized forcolorant and dyeing purposes.

Aqueous extract from *Tegetes* has shades of light yellowish green hue color. A laboratory experiment (Mihalick and Donnelly, 2006) showed that use of metal salts (mordants) can produce different colors on fabrics dyed with its flower extract. This seems to demonstrate the potentiality of colorant for fabric dyeing. As the flower has good pigment content *Tegetes erecta* flowers were chosen for the study of natural dyeing (Vankar, 2009).

1D.4.1 Traditional knowledge of its use

Ethanolic extract of the yellow to orangish red flowers of *Tegetes*, have been known as a rich source of lutein. This pigment has acquired greater significance because of its antioxidant property and for its use in eye health protection. Although *Tegetes* flower extract has been mainly used in veterinary feeds and as a natural food colorant. The chemical processing and stability of the pigment and its applications in these areas has been well established, but it has not been explored for fabric dyeing on commercial scale because of some drawbacks. The aqueous extract of the flower does not produce attractive hue color. Color fastnesses of these dyed fabrics do not fit into the acceptable norms of natural dyeing.

Since proper extraction of the colorant seemed to be the major drawback, we planned our study with *Tegetes* by developing an efficient extraction process using ethanol. The innovative extraction method gave better dye yield and deeper hue color to the extract. The ethanol was removed under reduced pressure on a rotatory evaporator; aqueous extract was prepared and used for natural dyeing of the three natural materials.

Although some attempts on natural dyeing with *Tegetes* has been made by natural dyers jute dyeing has been attempted by many researchers (Pan et al., 2003; Pan et al., 2004) and as a dye in silk dyeing (Mahale et al., 1999). Excellent colors were produced on sheepskin leather (Karolia and Dilliwar, 2004) with this dye when it was used in conjunction with three natural tannins and three mordants. Fabric printing and its effect with seven mordants through simultaneous mordanting on cotton and other fabric has been studied (Agarwal et al., 2007).

1D.5 *Nephelium lappaceum* (Rambutan) pericarp

Plant: *Nephelium lappaceum*
Family: Sapindaceae
Genus: Nephelium
Part used: Dried Fruit pericarp

Fig. 1D.5 Dried fruit pericarp of Rambutan.

Nephelium lappaceum (rambutan) is originally from Malaysia and Indonesia, but is now cultivated throughout the tropics particularly in Srilanka (Fig. 1D.5). Commercial production is primarily concentrated in Thailand, Malaysia, Indonesia, Philippines, Australia, Sri Lanka, Vietnam, Honduras, and Hawaii (Morton, 1987). The rambutan tree reaches 50–80 ft (15–25 m) in height, has a straight trunk to 2 ft (60 cm) wide, and a dense, usually spreading crown. The fruit is ovoid, or ellipsoid, pinkish-red, bright-or deep-red, orange-red, maroon or dark-purple, yellowish-red, or all yellow, or orange-yellow; 1 1/3–3 1/8 in. (3.4–8 cm) long. Its thin, leathery rind is covered with tubercles from each of which extends a soft, fleshy, red, pinkish, or yellow spine 1/5–3/4 in. (0.5–2 cm) long, and the tips deciduous in some types. The somewhat hair like covering is responsible for the common name of the fruit, which is based on the Malay word *rambut*, meaning "hair." It occurs in humid valley forest of Srilanka. The Rambutan flourishes from sea-level to 1600 or even 1800 ft (500–600 m), in tropical, humid regions having well-distributed rainfall.[4]

Rambutan fruits have been used as natural dye-stuff in indigenous systems which has been developed scientifically and can be substituted for the chemical dyes for local handloom sector for shades of light brown-dark brown as shown in cotton, silk, and wool dyed samples.

[4] https://www.hort.purdue.edu/newcrop/morton/rambutan.html.

Nephelium lappaceum (Rambutan) belongs to family Sapindaceae. It produces natural dye from the pericarp of its fruit, which has been used for dyeing textiles. In the present study innovative dyeing with Rambutan has been shown to give good dyeing results. Pretreatment with 2% metal mordant and using 5% of plant extract (owf) is found to be optimum and shows very good fastness properties for cotton, wool, and silk dyed fabrics (Vankar et al., 2007).

1D.6 *Curcuma domestica Valet* extract

Plant: *Curcuma domestica*
Order: Zingiberales
Family: Zingiberaceae
Genus: Curcuma
Part used: Dried Roots

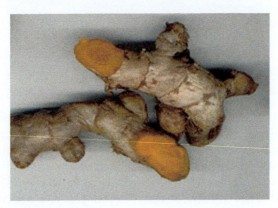

Fig. 1D.6 Fresh rhizome of curcuma.

Curcuma domestica Valet (Turmeric) is a common culinary spice used in Asian countries (Fig. 1D.6). It belongs to *Zingiberaceae* family. The tuberous rhizomes or underground stems of turmeric have been used from antiquity as condiment, a dye, and as an aromatic stimulant in several medicines. Turmeric is a very important spice in India; it is cultivated in Sri Lanka as well. Turmeric has been used in Indian systems of medicine- ayurveda, for a long time.

An investigation was carried out to find out the effect of sonicator dyeing with aqueous extract of *Curcuma domestica Valet* and to evaluate fastness properties by the extract on cotton, silk, and wool and was compared with conventionally dyed samples mainly with an aim to improve wash and light fastnesses of dyed fabrics (Vankar et al., 2008). From the study it was found that general appearance, luster, and texture of the sonicator dyed samples were better which showed marked difference in wash fastness, light fastness, and color strength C and H values.

1D.6.1 Dyeing properties of curcuma

The yellow constituent of curcuma, curcumin, has been utilized for dyeing wool and cotton fabrics. It is also employed as a coloring material in pharmacy, confectionery, food industries, and also in paints and varnishes (Boruah and Kalita 2015). Several workers have used curcuma as a source of natural yellow dyes particularly for cotton and silk. Curcuma was used for dyeing Eri silk after metal mordanting. Curcuma was also used for dyeing cotton, where the scientist showed that the fabric was dyed with four different natural dyes (turmeric, myrobolan, madder, red sandalwood) using pre, post, and simultaneous-mordanting techniques (Gogoi et al., 1997; Teli and Nayak, 1994). The application of single and mixture of selected natural dyes including the use of curcuma on cotton fabric is shown by a scientific approach (Samanta et al., 2003). Although the dyeing showed good dye uptake, the dye from curcuma was found to be very fugitive displaying poor wash and light fastnesses (Yoshizumi and Crews, 2003).

Fading behavior of natural yellow dyes and the maintenance of Japanese yellow kosode (dyes) (Ishii and Saito, 2006) was studied, the dyed silk and cotton fabrics were exposed to xenon-arc lamp and the fading rate was measured against Japanese Std. (JIS) blue scale and fluorescence scale. Reflectance spectrum, CIELAB, and color difference (δE) were gained and evaluated. Samples were photographed under day light and UV light for record where Turmeric showed very poor results. Characteristics of fading of wool cloth dyed with selected natural dyestuffs on the basis of solar radiant energy was investigated by researchers (Yoshizumi and Crews, 2003) where turmeric among other yellow dyes showed very fast fading.

Dyeing behavior and bacteria repellency of natural vegetable dye turmeric has also been studied (Lin et al., 2004). Antimicrobial activity of wool fabric treated with curcumin was attempted (Han and Yang, 2005). Effect of enzymes and protein (casein) has also been attempted with turmeric (Tsatsaroni et al., 1998) in order to improve its dye ability.

Investigation of the effect of sonicator in natural dyeing with aqueous extract of *Curcuma domestica Valet* and to evaluate fastness properties by the extract on cotton, silk, and wool and compared these samples with conventionally dyed samples mainly with an aim to improve wash and light fastnesses of dyed fabrics.

Several other flowers, barks, and leaves such as Bougainvellia (Tiwari et al., 2000d), Cineneria (Tiwari et al., 2000a), Balsam (Tiwari et al., 2000c), Eucalyptus (Tiwari et al., 2000b), Sappan wood (Ghorpade et al., 2000), and *Cassia fistula* (Tiwari et al., 2001), were investigated. Most of these dyes gave good dyeing performance when judged by internationally accepted color and dyeing standards.

References

Agarwal, R., Pruthi, N., Singh, S.J.S., 2007. Effect of mordants on printing with Marigold flowers dye. Nat. Prod, Rad. 6, 306–309.

Boruah, S., Kalita, B., 2015. Eco-printing of Eri silk with turmeric natural dye. Int. J. Text. Fashion Technol. 5, 27–32.

Bosma, T., Dole, J., Maness, N., 2000. 621 Optimizing marigold (*Tagetes erecta* L.) petal and pigment yield. HortSci. 35, 504.

Ferenczi, S., Kallay, M., Bardi, G., 1981. Natural dyes preparations from Hollyhock petals. Teljes HU 26, 789.

Ghorpade, B., Darvekar, M., Vankar, P.S., 2000. Ecofriendly cotton dyeing with Sappan wood dye using ultrasound energy. Colourage 27.

Gogoi, A., Ahmed, S., Barua, N., 1997. Natural dyes & silk. Indian Text. J. 107, 64–73.

Han, S., Yang, Y., 2005. Antimicrobial activity of wool fabric treated with curcumin. Dyes Pigments 64, 157–161.

Ishii, M., Saito, M., 2006. Fading behavior of natural yellow dyes and the maintenance of Japanese yellow kosode [Original title and text in Japanese]. Bunkazai hozon-syuhuku gakkaisi 51, 14–37.

Karolia, A., Dilliwar, S., 2004. Natural yellow dyes from marigold flowers for leather. Colourage 51, 31–38.

Kasumov, M., 1984. Red dye from the hollyhock and its use in food industry. Dokl. Akad. Nauk Az. SSR 40, 76–79.

Lin, M.-X., Wu, J., Shi, M., 2004. Dyeing behaviour and bacteria repellency of natural vegetable dye turmeric. J. Dalian Inst. Light Ind. 23, 59–62.

Mahale, G., Bhavani, K., Sunanda, R., 1999. Standardising dyeing conditions for African Marigold. Man Made Text. India 42, 453–458.

Matula, V., Macek, C., 1936. The anthocyanins as indicators in neutralization analysis. Chemicky Obzor 11, 83–84.

Mihalick, J.E., Donnelly, K.M., 2006. Using metals to change the colors of natural dyes. J. Chem. Educ. 83, 1550.

Morton, J.F., 1987. Rambutan. JF Morton, Miami, FL.

Pan, N., Chattopadhyay, S., Day, A., 2003. Dyeing of jute with natural dyes. Ind. J. Fibre Text. Res. 28, 339–342.

Pan, N.C., Chattopadhyay, S.N., Day, A., 2004. Dyeing of jute fabric with natural dye extracted from marigold flower. Asian Text. J. 13, 80–82.

Samanta, A., Singhee, D., Sethia, M., 2003. Application of single and mixture of selected natural dyes on cotton fabric: a scientific approach. Colourage 50, 29–42.

Shanker, R., Vankar, P.S., 2007. Dyeing cotton, wool and silk with Hibiscus mutabilis (Gulzuba). Dyes Pigments 74, 464–469.

Subramanian, S.S., Nair, A., 1970. Sterols and flavonols of *Ficus bengalensis*. Phytochemistry 9, 2583–2584.

Teli, M.D., Nayak, A.N., 1994. Natural dyes—dyeing of cotton with turmeric. J. Text. Assoc. 7, 82.

Tiwari, V., Ghorpade, B., Mishra, A., Vankar, P.S., 2000a. Ultrasound energized dyeing of cotton fabric with Cineraria flowers using ecofriendly mordants. Asian Text. J. 58.

Tiwari, V., Ghorpade, B., Mishra, A., Vankar, P.S., 2000b. Ultrasound energised dyeing of cotton fabric with Eucalyptus bark. Asian Text. J. 30.

Tiwari, V., Ghorpade, B., Vankar, P., 2000c. Ultrasonic dyeing with impatiens balsamina (balsam) using ecofriendly mordants on cotton. Colourage 47, 21–22.

Tiwari, V., Ghorpade, B., Vankar, P.S., 2000d. Ultrasound energised dyeing with aqueous extract of Bougainvillea using ecofriendly mordants. Asian Text. J., 28.

Tiwari, V., Ghorpade, B., Mishra, A., Vankar, P.S., 2001. Ultrasonic energised dyeing of cotton with Cassia fistula bark (Amaltas). Asian Text. J. 56.

Tsatsaroni, E., Liakopoulou-Kyriakides, M., Eleftheriadis, I., 1998. Comparative study of dyeing properties of two yellow natural pigments—effect of enzymes and proteins. Dyes Pigments 37, 307–315.

Vankar, P.S., 2009. Utilization of temple waste flower—*Tagetus erecta* for dyeing of cotton, wool and silk on industrial scale. J. Text. Appar. Technol. Manag. 6.

Vankar, P.S., Shanker, R., 2006a. Dyeing cotton, silk and wool with *Cayratia carnosa* Gagn.or Vitis trifolia. Asian Text. J. 38.

Vankar, P.S., Shanker, R., 2006b. Dyeing silk, wool and cotton with *Alcea rosea* flower. Fibre 2 Fashion.

Vankar, P., De Alwisb, A., De Silvab, N., 2007. Dyeing of dyeing cotton, wool and silk with extract of Nephelium lappaceum (Rambutan) pericarp. Asian Text. J., 66–70.

Vankar, P., Shanker, R., Wijayapala, S., De Alwisb, A., De Silvab, N., 2008. Sonicator dyeing of cotton, silk and wool with *Curcuma domestica* Valet extract. Int. Dyers 193, 38–42.

Vasudevan, P., Kashyap, S., Sharma, S., 1997. Tagetes: a multipurpose plant. Bioresour. Technol. 62, 29–35.

Yoshizumi, K., Crews, P.C., 2003. Characteristics of fading of wool cloth dyed with selected natural dyestuffs on the basis of solar radiant energy. Dyes Pigments 58, 197–204.

Isolation and characterization of the colorant molecules from each dye plant

2

P.S. Vankar
FEAT (Facility for Ecological and Analytical Testing), Kanpur Kalyanpur, India

Introduction

Dye compounds from natural resources especially from plants are increasingly becoming important alternatives to synthetic dyes for use in the textile industry (Deo and Desai, 1999; Samanta et al., 2009). The perception of color is an ability of some animals, including humans, to detect some wavelengths of electromagnetic radiation (light) differently from other wavelengths. Dyes possess color because they

- absorb light in the visible spectrum (400–700 nm);
- have at least one chromosphores (color-bearing group);
- have a conjugated system, i.e., a structure with alternating double and single bonds;
- exhibit resonance of electrons, which is a stabilizing force in organic compounds (Abrahart, 1977).

When any one of these features is lacking from the molecular structure, the color is lost. In addition to chromophores, most dyes also contain groups known as auxo-chromes (color helpers), examples of which are carboxylic acid, sulfonic acid, amino group, and hydroxyl group. The word auxochrome is derived from two roots. The prefix auxo is from auxein, which means increased. The second part, chrome, means color, so the basic meaning of the word auxochrome is color increaser. This word was coined because it was noted originally that the addition of ionizing groups resulted in a deepening and intensifying of the color of compounds. While these are not responsible for color, their presence can shift the color of a colorant, and they are most often used to influence dye solubility. In order to know the structure of the colorant in each dye extract that was derived from different plants, it was first necessary to separate the col-ored molecules by column chromatography. The structure elucidation of separated and isolated compounds was then identified through spectroscopic methods such as UV-visible, FT-IR, and mass spectroscopic methods. Although these are tedious methods, they help in matching the structures with phytochemical literature and data.

The presence of colorant and its chemical nature was ascertained by spectroscopic and chromatographic analysis of the dye extract. *UV-visible* analysis of aqueous and/or methanolic extract was carried out on Thermo Heλios α model spectrophotometer at a resolution of 1 nm. *FT-IR* analysis of methanolic extract was carried out on by Vertex 70 model of Bruker. *HPLC* was taken generally in methanol-water (60:40) system on C18 column with flow rate of 1 mL on Waters HPLC.

Natural Dyes for Textiles. http://dx.doi.org/10.1016/B978-0-08-101274-1.00002-1

2.1   Chemical characterization of the acacia bark extract

The acacia bark contains 12%–20% of tannins, with a tannin/nontannin ratio of 1:5, and contains catechin, epicatechin, dicatechin, quercetin, and procyanidin. Other compounds present such as acacidiol, resorcinol, methyl gallate, methyl 5-*O*-methylgallate, and ecosanamide have been isolated from the stems of *Acacia arabica* (Singh and Kalidhar, 2004). The tannins, which belong to the proanthocyanidin type, increase in content in the bark as the trees age. The acacia bark primarily consists of tannins of two types—hydrolyzable tannins and proanthocyanidins. The hydrolyzable tannins are molecules with polyol (generally D-glucose) as a core. The hydroxyl groups of these carbohydrates partially or totally esterify with phenolic groups like gallic acid or ellagic acid as shown in Fig. 2.1. Proanthocyanidins are present in larger quantities and have flavonoid units with different substituent. These anthocyanidin pigments are deeply colored due to the presence of free hydroxyl moiety as shown in Fig. 2.2.

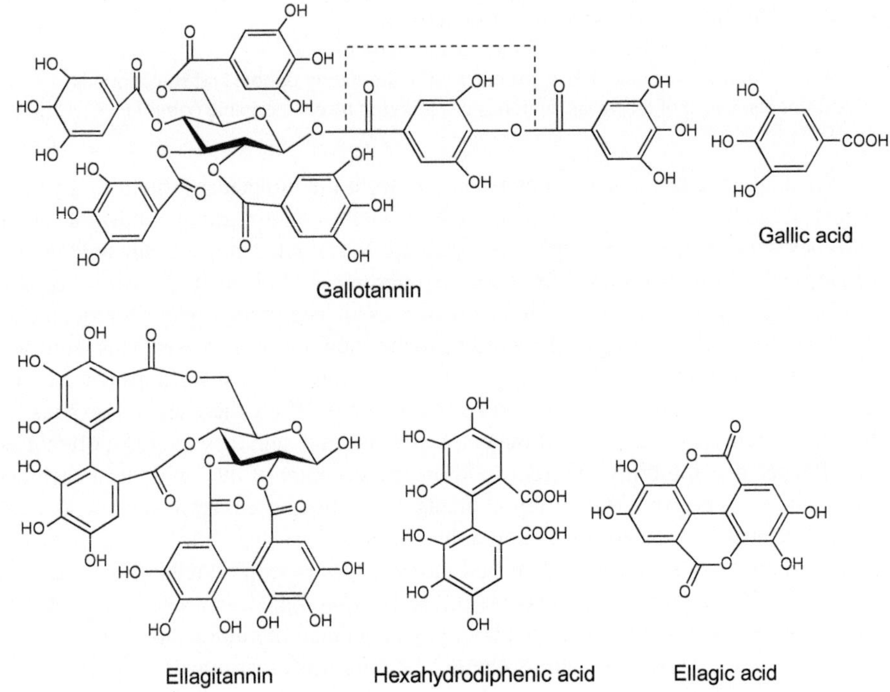

Gallotannin

Gallic acid

Ellagitannin   Hexahydrodiphenic acid   Ellagic acid

Fig. 2.1 Components of acacia bark.

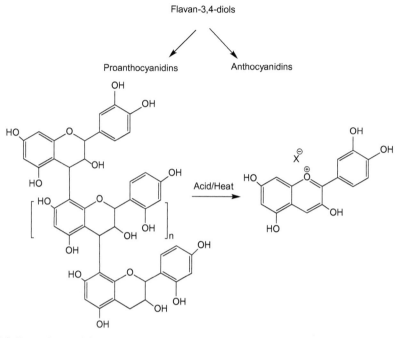

Fig. 2.2 Proanthocyanidins and anthocyanidin with flavonoid units in acacia bark.

2.1.1   HPLC of the acacia bark extract at different wavelengths

HPLC of acacia bark was carried out with solvent system 95:5 MeOH-deionized water (DW) having time of 20 min, and sample was prepared in MeOH, chromatogram taken at 255 nm and at 280 nm (Figs. 2.3 and 2.4).

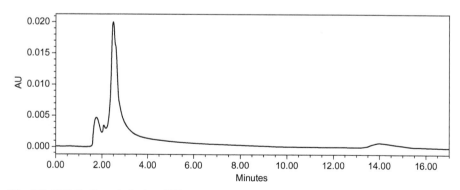

Fig. 2.3 HPLC of acacia bark at 255 nm.

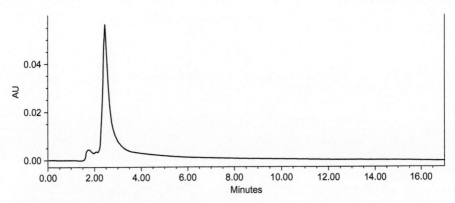

Fig. 2.4 HPLC of acacia bark at 280 nm.

2.2 Chemical characterization of the Mahonia stem extract

The methanolic extract from *Mahonia napaulensis* gave five fractions by column chromatography. Each of the color components isolated from stem extract of *M. napaulensis* that showed a characteristic visible spectrum from which a λ_{max} of 410 nm was considered for chemical characterization. Figs. 2.5 and 2.6 show the structures of berberine and ampelopsin (flavanol) isolated from the extract.

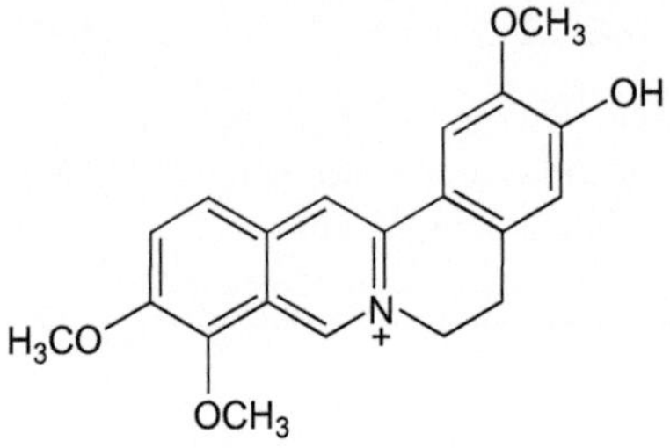

Fig. 2.5 Berberine.

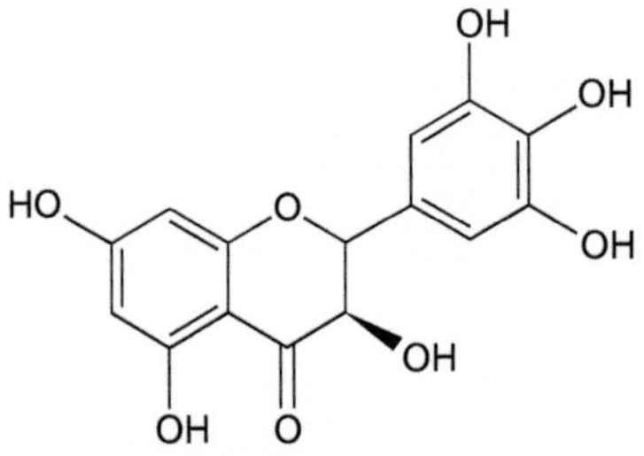

Fig. 2.6 Ampelopsin.

2.3 Chemical characterization of the colorant Salvia flower extract

Salvianin, which was pelargonidin diglucoside, was isolated for the first time by Willstätter et al. (1917), which was followed by Tomás-Barberán et al. (1987) who isolated pelagonidin 3-caffeylglucoside-5-dimalonylglucoside and monardaein. The complex structure of the dye is shown below, which was described in Kondo et al. (1989) (Figs. 2.7 and 2.8).

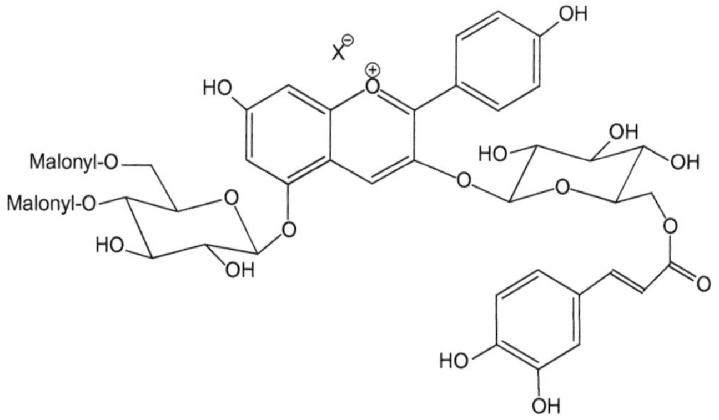

Fig. 2.7 Structure of pelagonidin 3-caffeylglucoside-5-dimalonylglucoside.

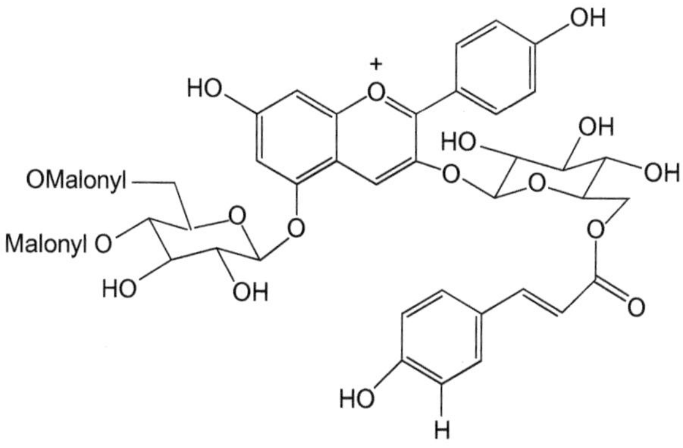

Fig. 2.8 Structure of monardaein.

2.4   Chemical characterization of Canna flower extract

Pigments from red *Canna* flowers have been isolated and identified as novel anthocyanins. They are cyanidin derivatives cyanidin-3-*O*-(6″-*O*-α-rhamnopyranosyl)-β-glucopyranoside) (Fig.2.9), cyanidin-3-*O*-(6″-*O*-α-rhamnopyranosyl)-

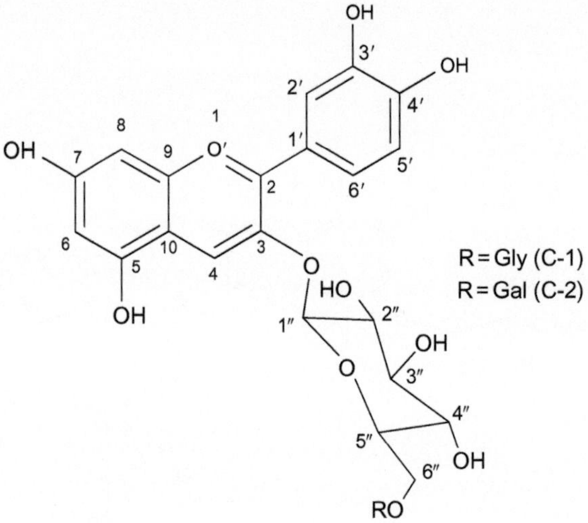

Fig. 2.9 Anthocyanin from *Canna* cyanidin-3-*O*-(6″-*O*-α-rhamnopyranosyl)-β-glucopyranoside. R = Glucopyranoside.

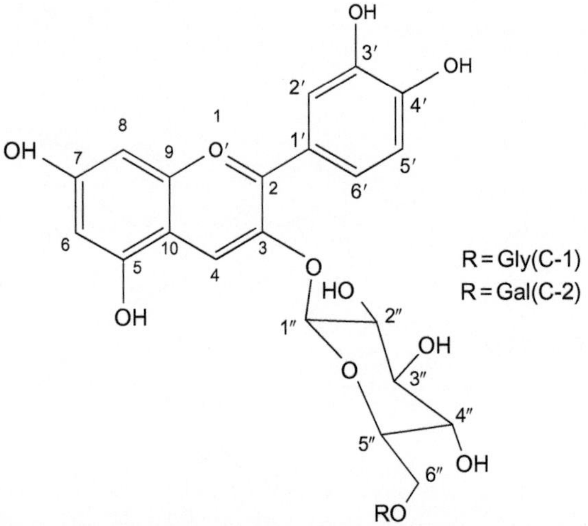

Fig. 2.10 Anthocyanin from *Canna* cyanidin-3-O-(6″-*O*-α-rhamnopyranosyl)-β-galactopyranoside same structure with R = Galactopyranoside.

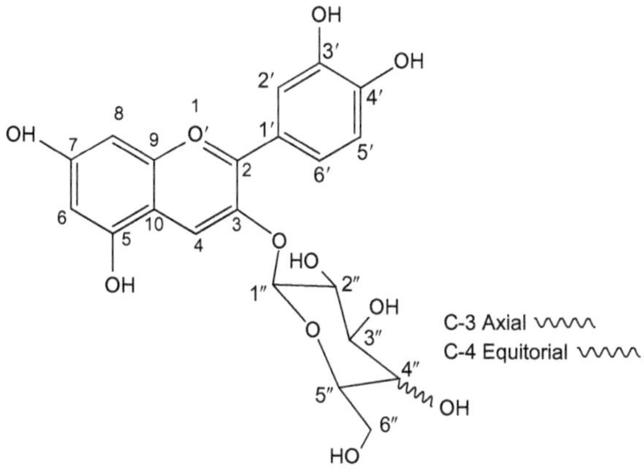

Fig. 2.11 Cyanidin derivative (cyanidin-3-*O*-β-glucopyranoside). C-3 = Axial.

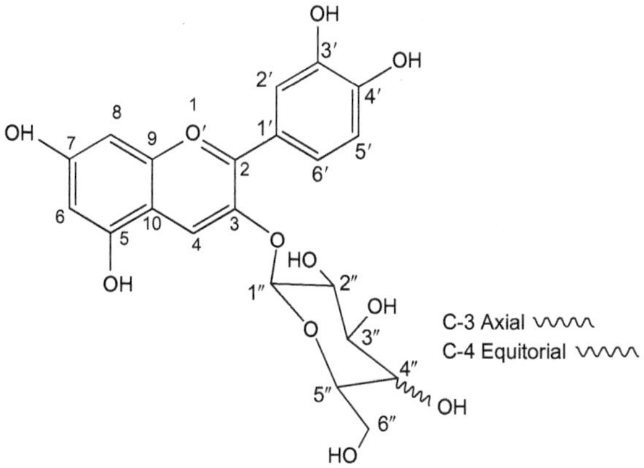

Fig. 2.12 Cyanidin derivative (cyanidin-*O*-β-galactopyranoside). C-4 = Equitorial.

β-galactopyranoside (Fig. 2.10), cyanidin-3-*O*-β-glucopyranoside (Fig. 2.11), and cyanidin-*O*-β-galactopyranoside (Fig. 2.12).

2.4.1 UV/visible spectrum of Canna

The UV-visible spectra of canna flower extract was taken at different conditions such as aqueous, acidic, and alkaline as shown in Fig. 2.13A–C. The λ_{max} is very different in each case.

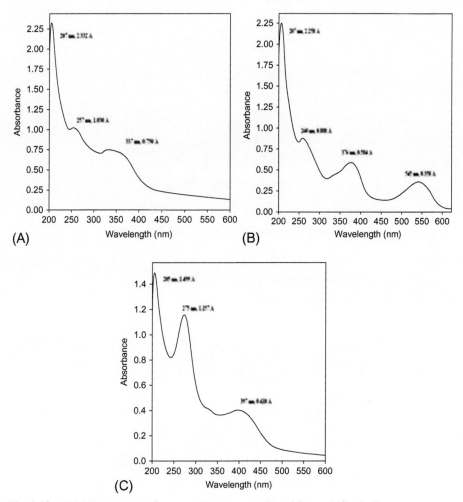

Fig. 2.13 UV/visible spectra of extracts (A) aqueous, (B) acidic, and (C) alkaline.

2.4.2 HPLC of Canna flower extract

The mobile phase consisted of A and B solvent systems. A was 100% HPLC grade acetonitrile, and B was 1% formic acid and 99% methanol. The program followed a linear gradient from 0% to 30% A in 30 min. Sample was prepared in methanol. Chromatograph was taken at 520 nm (Fig. 2.14).

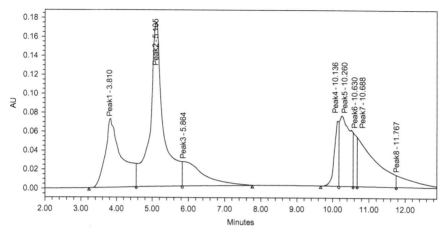

Fig. 2.14 HPLC of *Canna* extract.

2.5 Chemical characterization of Rhododendron flower extract

The flower extract of red variety of *Rhododendron* has been identified as flavonols, flavonol glycosides, flavanes, and dihydroflavonols such as afzelin, ampelopsin, catechin, myricetin, myricitrin, quercetin, and quercitrin. Chemical composition in flower extract of *Rhododendron dauricum* (Jung et al., 2007) was shown by the presence of total flavonoids in a content of 30%–90% hyperin 6.3%–42%, farrerol 0.4%–8%, and quercetin 0.7%–7% (Figs. 2.15–2.17). Dark-red variety of *Rhododendron* flowers aq. extract showed one major peak, λ_{max} at 535 nm in the visible region as shown in Fig. 2.18.

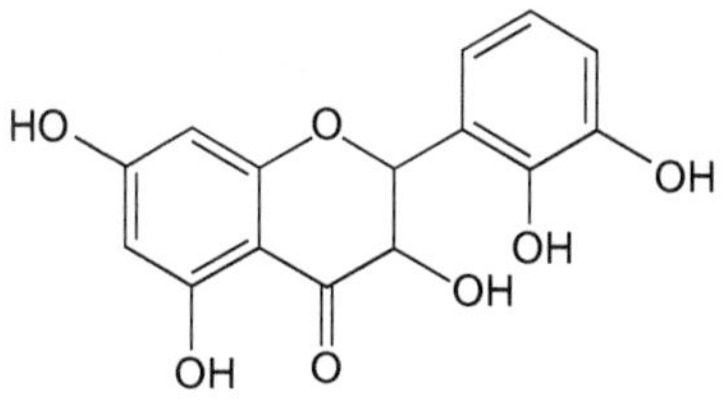

Fig. 2.15 Quercetin.

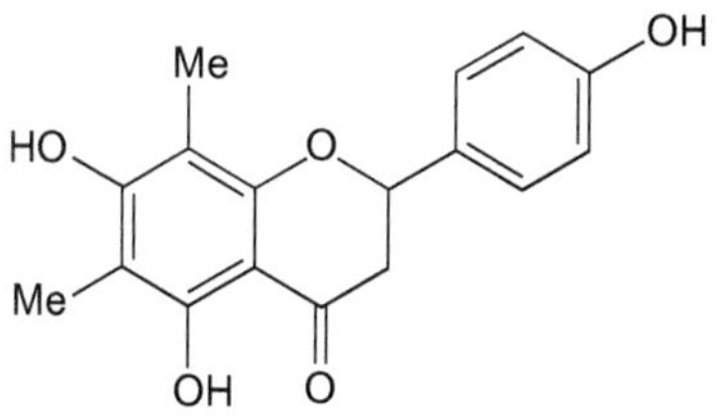

Fig. 2.16 Farrerol.

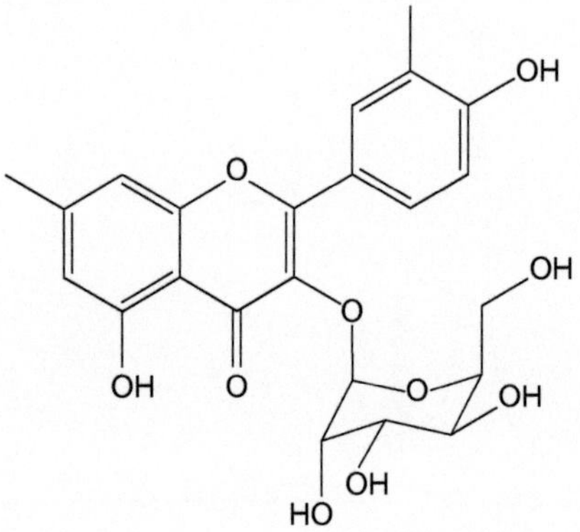

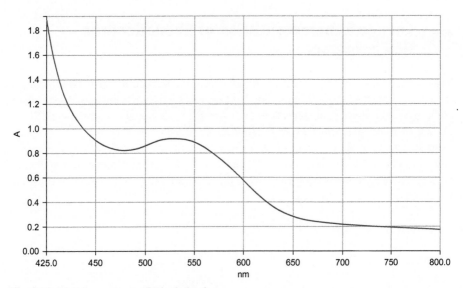

Fig. 2.17 Hyperin.

Fig. 2.18 Visible spectrum of *Rhododendron*.

2.6   Chemical characterization Cosmos flower

The main coloring agent in *Cosmos sulphureus* is a pentahydroxy chalcone hexo-side, an anthochlorine type flavonoid generically known as coreopsin as shown in Fig. 2.19. However, coloring is also provided by additional flavonoids in the plant, such as isoquercetin and the luteolin glycosides (golden yellow). Chalcones constitute a minor family of substances belonging to the flavonoids. Sulfuretin, an aurone was also found to be present in the flower extract of cosmos.

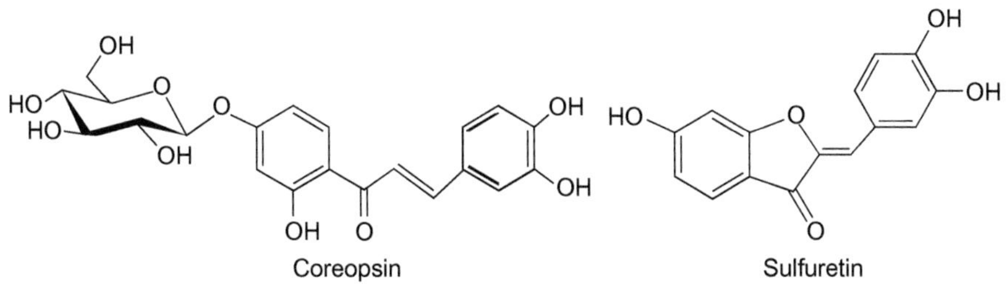

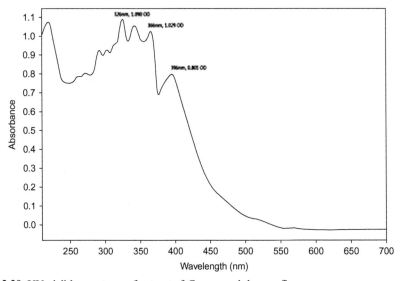

Coreopsin   Sulfuretin

Fig. 2.19 Coreopsin and sulfuretin.

2.6.1   Ultraviolet & visible spectra of Cosmos

See Fig. 2.20.

Fig. 2.20 UV-visible spectrum of extract of *Cosmos sulphureus* flowers.

2.6.2   FT-IR spectrum of cosmos

See Fig. 2.21.

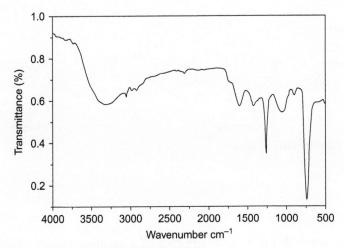

Fig. 2.21 FT-IR of extract of *Cosmos sulphureus* flowers.

2.6.3 HPLC of Cosmos flower extract

Solvent system: 95:5 MeOH-H$_2$O
   Run time: 15 min
   The sample was prepared in MeOH; chromatogram was taken at 255 nm (a) and at 280 nm (b) (Figs. 2.22 and 2.23).

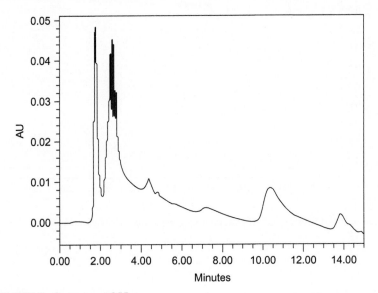

Fig. 2.22 HPLC of cosmos at 255 nm.

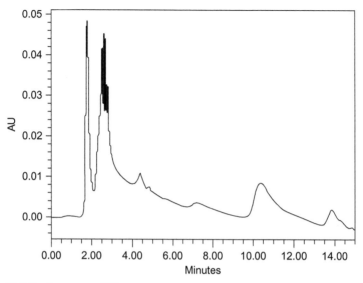

Fig. 2.23 HPLC of cosmos at 280 nm.

2.7 Chemical composition of *Terminalia arjuna* extract

An alcoholic extract of the bark yielded a red amorphous coloring matter, arjunetin; it is a glucoside, $C_{11}H_{18}O_4 \cdot H_2O$, having a melting point of 215°C. Other compounds present are cerasidin, β-sitosterol, friedeline, methyl oleanolate, and gallic, ellagic, and arjunic acids (Fig. 2.24).

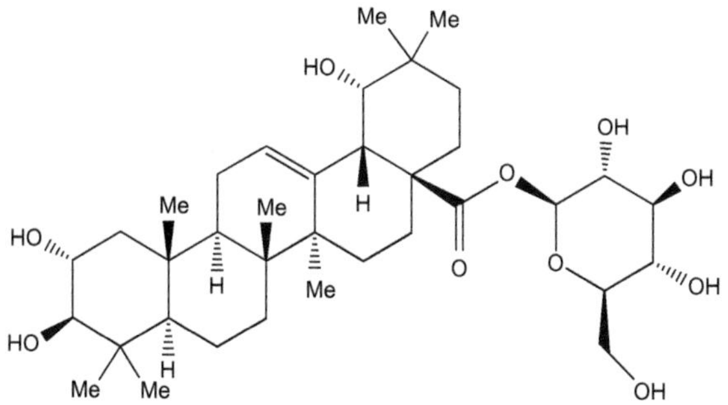

Fig. 2.24 Arjunetin.

2.7.1   Ultraviolet & visible spectra

See Fig. 2.25.

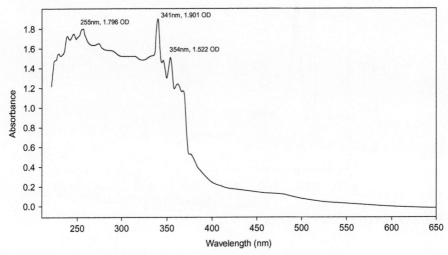

Fig. 2.25 UV-visible spectra of aqueous extract of *Terminalia arjuna* bark.

2.7.2   FT-IR spectrum

See Fig. 2.26.

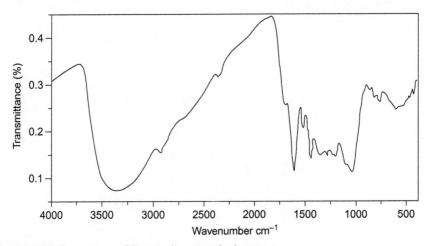

Fig. 2.26 FT-IR spectrum of *Terminalia arjuna* bark extract.

2.7.3 HPLC of Terminalia arjuna *bark extract*

Solvent system: 95:5 MeOH-DW
  Time: 15 min
  The sample was prepared in MeOH; chromatogram was taken at 255 nm and at 280 nm (Figs. 2.27 and 2.28).

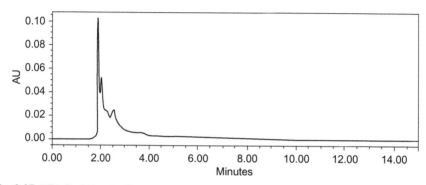

Fig. 2.27 HPLC of *Terminalia* extract at 255 nm.

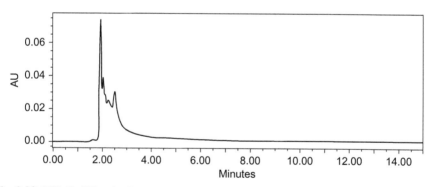

Fig. 2.28 HPLC of *Terminalia* extract at 280 nm.

2.8 Chemical characterization of black carrot

The visible spectrum of black carrot extract shows peaks at the region 525–660 nm as shown in Fig. 2.29. Three apparent peaks were seen in this region indicating the presence of cyanidin 3-xylosyl (sinapoyl-glucosyl)galactoside, cyanidin 3-xylosyl (feruloylglucosyl)galactoside, and cyanidin 3-xylosyl (coumaroylglucosyl)galactoside as isolated (Montilla et al., 2011).

2.8.1   UV-visible spectra of black carrot

See Fig. 2.29.

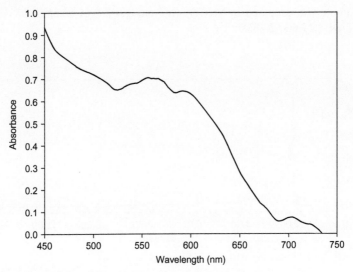

Fig. 2.29 Visible spectrum of the aqueous extract of the black carrot.

2.8.2   FT-IR spectra of black carrot

See Fig. 2.30.

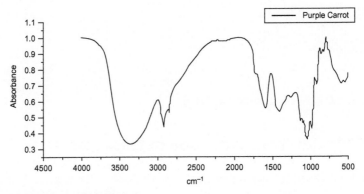

Fig. 2.30 FT-IR spectra of black carrot.

2.8.3 HPLC spectra of black carrot

The HPLC of the black carrot extracts showed the separation of anthocyanins as presented in Fig. 2.31. As shown, an efficient separation of five major anthocyanins and a few of minor compounds could be achieved within 30 min. They were assumed to be peonidin, cyanidin, and pelargonidin glycosides.

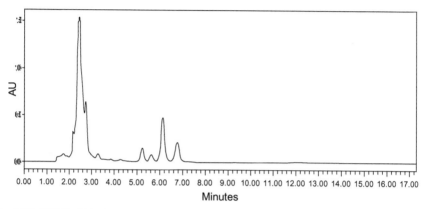

Fig. 2.31 HPLC of black carrot.

2.9 Chemical characterization of *Hibiscus rosa sinensis*

The *Hibiscus* anthocyanin mainly comprises of cyanidin-3-sophoroside as shown in Fig. 2.32.

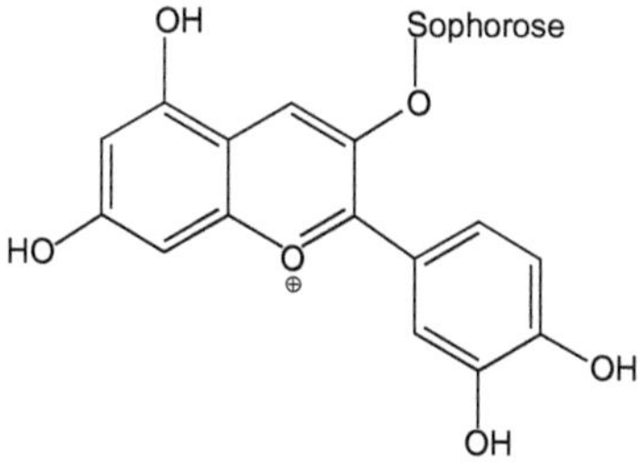

Fig. 2.32 Main colorant in hibiscus flower cyanidin-3-sophoroside.

2.9.1 HPLC of Hibiscus flower

Solvent system: 95:5 MeOH-H$_2$O
   Time: 25 min
   The sample was prepared in MeOH; chromatogram was taken at 255 nm (Figs. 2.33 and 2.34).

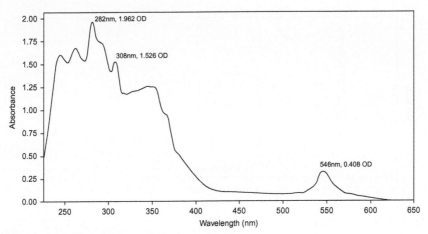

Fig. 2.33 UV-visible spectrum of hibiscus flower.

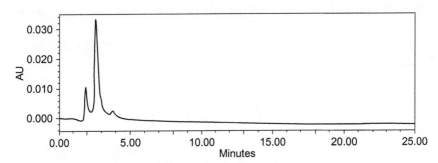

Fig. 2.34 HPLC of Hibiscus flower.

2.10   Chemical characterization of *Delonix regia* flower

The red color in *Delonix regia* flower is most probably due to copigmentation between anthocyanins and other flavonoids (Purohit et al., 2007); the color of the reddish-yellow flowers is mostly attributed to an increase in the isosalipurposide concentration, along with an increase in the background of the yellowish cytoplasmic carotenoids. A comparative study of the carotenoids present in various floral parts of *Delonix regia* has been made to gain information about the biogenesis and role of carotenoids in the flower. The qualitative and quantitative distribution of carotenoids was studied by chromatographic, spectrophotometric, and other methods. The partition ratios, hitherto not reported, of a number of different carotenoids between different solvents are reported. The petals contain 29 carotenoids. The major pigments found were phytoene, phytofluene, β-carotene, γ-carotene, lycopene isomers, rubixanthin, lutein, zeaxanthin, and several epoxy carotenoids (Gupta and Chandra, 1971).

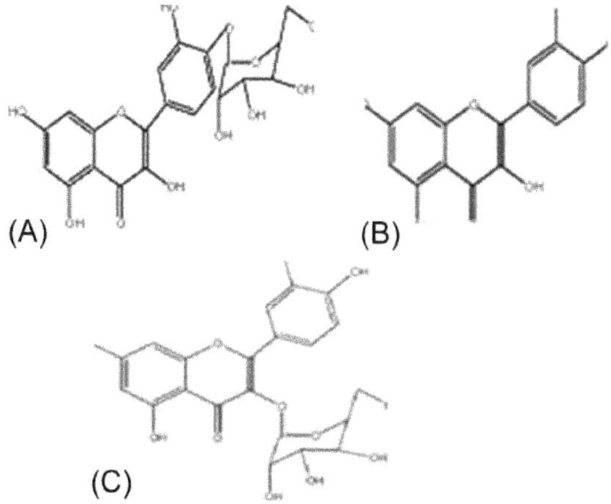

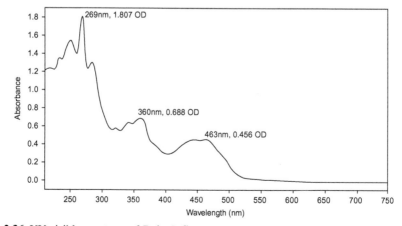

Fig. 2.35 Quercetin (B) and its different glucosides (A, C).

Beside the anthocyanins of *Delonix regia*, other flavonoids present were identified, namely, quercetin-3-rutinoside, quercetin-4′-glucoside, quercetin-3-glucoside (as shown in Fig. 2.35), quercetin-5-glucoside, and chalcononaringenin-2′-glucoside (Jungalwala and Cama, 1962). These pigments play an important role in dyeing of the fabric. Their chelation to biomordant or to enzyme would be a determining factor on their dyeability.

2.10.1 Ultraviolet & visible spectra of Delonix regia *flower* extract

See Fig. 2.36.

Fig. 2.36 UV-visible spectrum of *Delonix* flower.

2.10.2   HPLC of Delonix regia *flower extract*

Solvent system: 95:5 MeOH-DW

Run time: 20 min

The sample was prepared in MeOH; chromatogram was taken at 255 nm (a) and 280 nm (b) (Fig. 2.37).

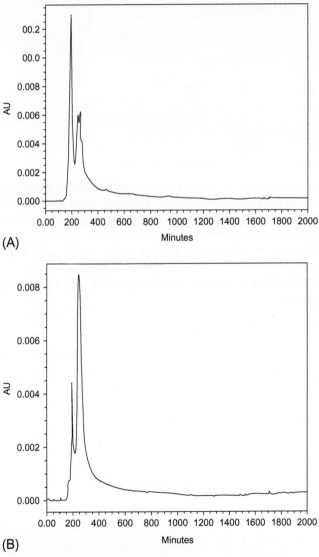

Fig. 2.37 (A) HPLC at 255 nm, (B) HPLC at 280 nm.

2.11   Chemical characterization of Plumeria flower

Petals of pink plumeria (0.25 kg) were extracted twice with water (1.5 L) at 50–55°C for 20–25 min, the extract was filtered, and slowly, the water is evaporated over water bath to get semidry paste (60 g) containing 12%–13% reddish-pink pigment. The flowers contain rutin-flavone, 3,3′,4′,5,5′,7-hexahydroxy-(6-*O*-α-L-rhamnosyl-β-D-glucoside), as shown in Fig. 2.38 and quercitrin (Fig. 2.39) as the main colorants.

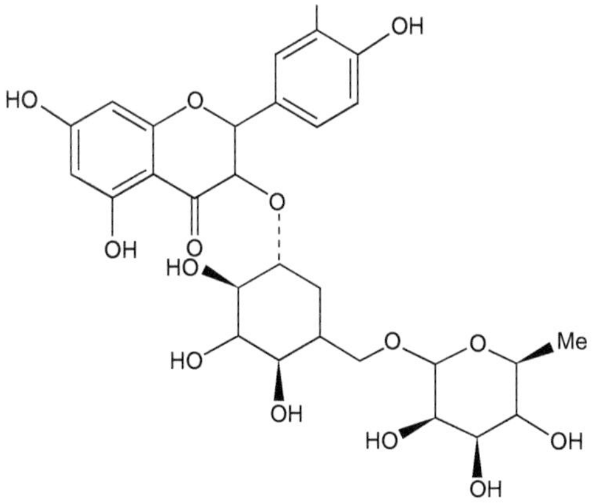

Fig. 2.38 Rutin.

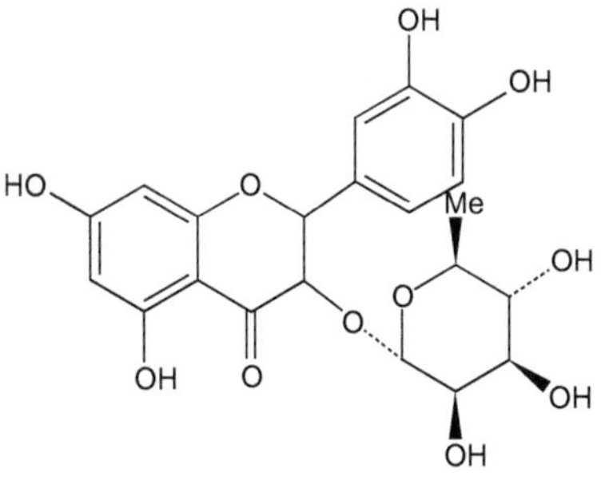

Fig. 2.39 Quercitrin.

2.11.1   Ultraviolet & visible spectra of Plumeria flower

The colorant showed one major peak, λ_{max} at 296.33 nm in the UV region (flavonoids) and at 547.96 nm in the visible region. The extract shows changes in visible graph at different pH as shown in Fig. 2.40. At pH 4, the λ_{max} was at 525 nm, 580 nm (pH 7), and 600 nm (pH 9).

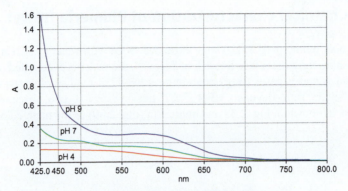

Fig. 2.40 Visible spectra of pink plumeria flower at different pH.

2.11.2 HPLC of Plumeria flower

Solvent system: 95:5 MeOH-DW

  Run time: 15 min

  The sample was prepared in MeOH; chromatogram was taken at 255 nm and at 280 nm (Figs. 2.41 and 2.42).

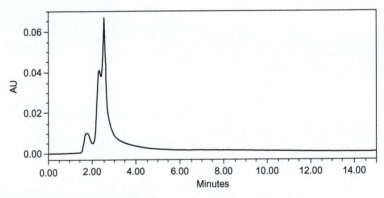

Fig. 2.41 HPLC of *Plumeria* at 255 nm.

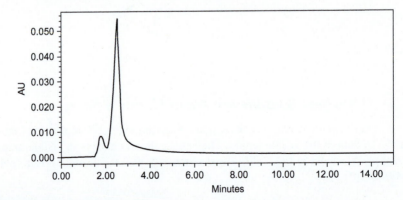

Fig. 2.42 HPLC of *Plumeria* at 280 nm.

2.12 Chemical characterization of *Combretum* (madhumalti) flower

The aqueous and methanolic extracts of the *Combretum* flower (madhumalti) shows slight change in visible spectrum of the madhumalti extract as shown in Fig. 2.43. The aqueous extract showed a λ_{max} at 538 nm with OD as 0.926 A, while the methanolic extract showed λ_{max} at 520 nm with OD as 0.807 A.

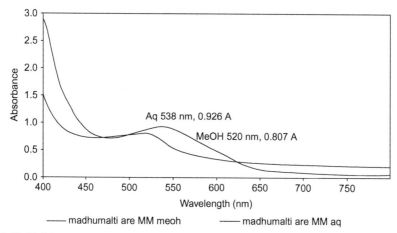

Fig. 2.43 Visible spectrum of madhumalti flower.

2.12.1 HPLC of madhumalti flowers

Solvent system: 95:5 MeOH-DW
    Run time: 15 min
    The sample was prepared in MeOH; chromatogram was taken at 255 nm and at 280 nm (Figs. 2.44 and 2.45).

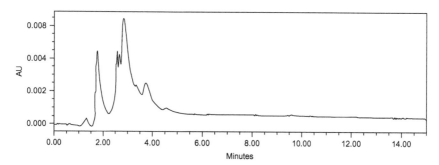

Fig. 2.44 HPLC of madhumalti flower at 255 nm.

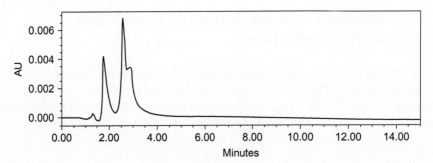

Fig. 2.45 HPLC of madhumalti flower at 280 nm.

2.13 Chemical characterization of the Ixora flower extract

Petals of *Ixora* (0.5 kg) were extracted twice with water (1.5 L) at 60–65°C for 20–25 min, the extract was filtered, and slowly, the water is evaporated on water bath to give dry-paste-like product (40 g) containing 7%–8% reddish pigment. Preliminary analysis showed that the pigment contained the phytochemical investigation of *Ixora coccinea* led to the isolation and identification of chrysin 5-O-β-D-xylopyranoside and flavone 4′,5,7-trihydroxy-, 4′-β-D-glucopyranoside along with apigenin 4′-O-β-D-glucopyranoside (Latha et al., 2001b; Latha and Panikkar, 1998; Latha et al., 2001a). Other compounds have been also isolated from *Ixora* (Ragasa et al., 2004) as shown in Figs. 2.46 and 2.47, respectively.

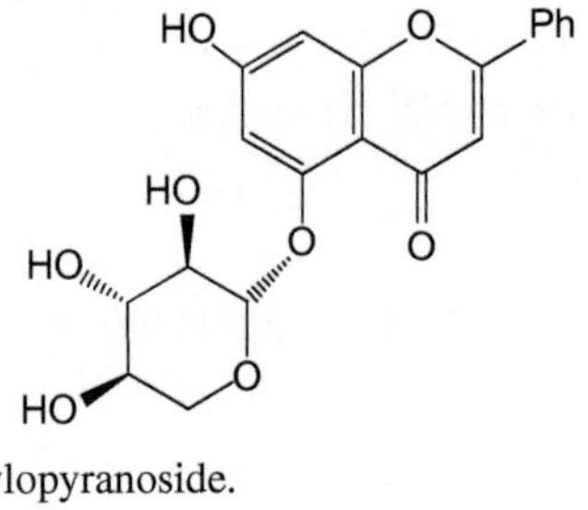

Fig. 2.46 Chrysin 5-O-β-D-xylopyranoside.

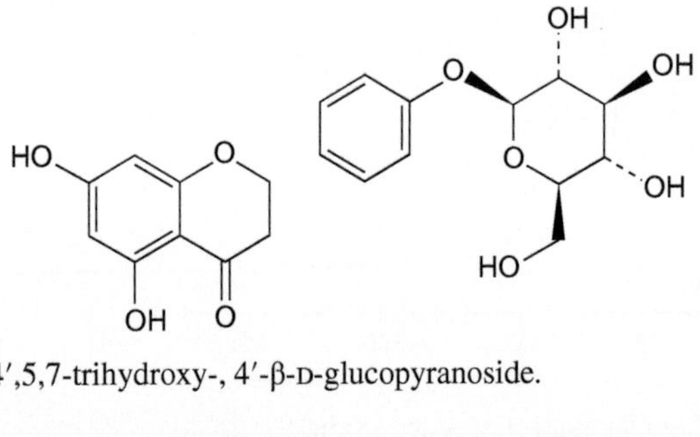

Fig. 2.47 Flavone 4′,5,7-trihydroxy-, 4′-β-D-glucopyranoside.

2.13.1 Ultraviolet & visible spectra of Ixora

See Figs. 2.48 and 2.49.

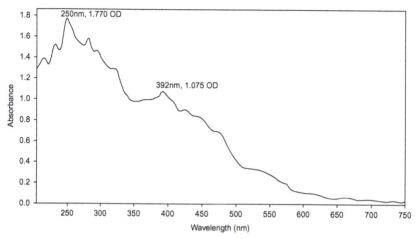

Fig. 2.48 Visible spectra of red *Ixora* flower extract.

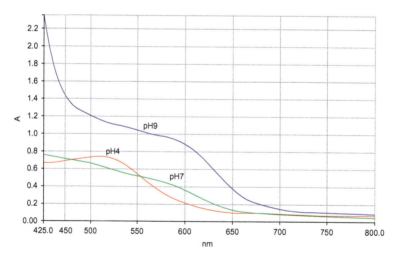

Fig. 2.49 Visible spectra of red *Ixora* flowers at different pH.

2.13.2 HPLC of Ixora flower extract

Solvent system: 95:5 MeOH-H$_2$O

Run time: 15 min

The sample was prepared in MeOH; chromatogram was taken at 255 nm (a) and 280 nm (b) (Figs. 2.50 and 2.51).

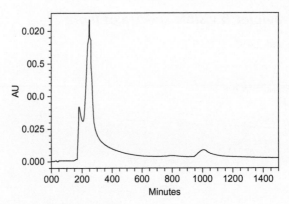

Fig. 2.50 HPLC of *Ixora* flower at 255 nm.

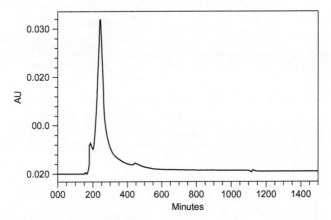

Fig. 2.51 HPLC of *Ixora* flower at 280 nm.

2.14   Chemical characterization of *Bischofia* extract

Bischofia javanica contains mainly the following compounds (Gupta et al., 1988; Tanaka et al., 1995) that were identified as n-triacontane, β-amyrin, friedeline, β-sitosterol, ursolic acid, chrysoeriol, fisetin, quercetin (as shown in Figs. 2.52–2.54),

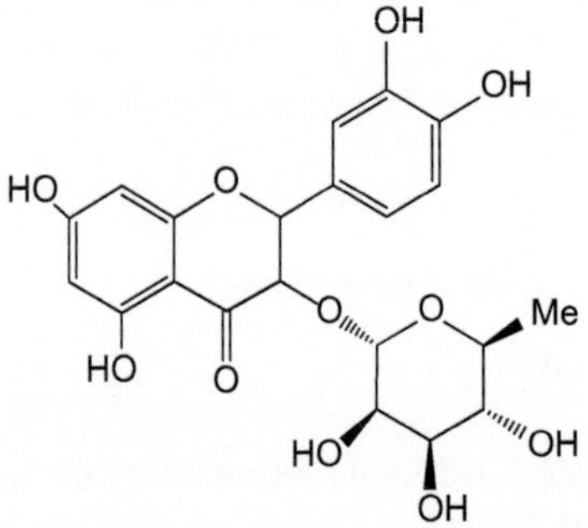

Fig. 2.52 Quercitrin.

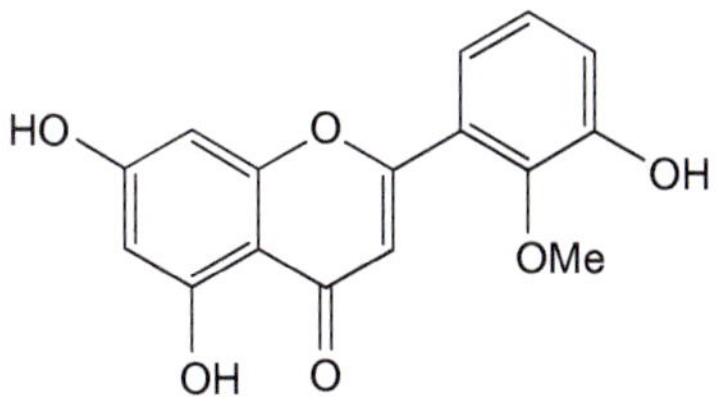

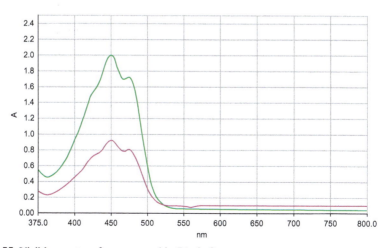

Fig. 2.53 3′-Methoxy-4′,5,7-trihydroxyflavone

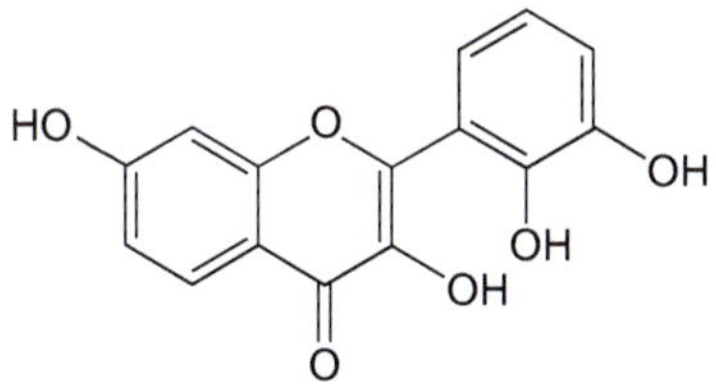

Fig. 2.54 Fisetin.

luteolin-7-O-glucoside, and quercitrin. Structures of some salient colorant molecules are given below. Column chromatography of the crude extract of *Bischofia* gave many colored compounds having colors like dark, yellow, orange, green, and brown confirming the presence of quercitrin. These colored compounds were then analyzed spectroscopically and identified (Vankar et al., 2007).

2.14.1 *Ultraviolet & visible spectra of Bischofia extract*

See Figs. 2.55 and 2.56.

Fig. 2.55 Visible spectra of a compound in *Bischofia*.

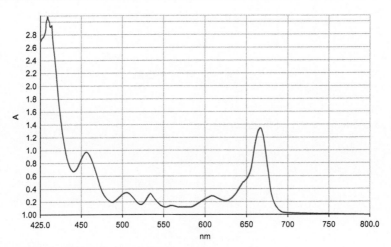

Fig. 2.56 Visible spectra of a compound in *Bischofia*.

2.15 Chemical characterization of *Celosia crisata* (Murgkesh)

Celosia cristata is a seasonal red flower, and it majorly contains amaranthine in its petals. Amaranthine, the characteristic pigment in the Amaranthaceae, was recently analyzed in extracts of red inflorescences of common cockscomb (*Celosia argentea* var. *cristata*). The identification was based on LC-MS of the purified compound having expected fraction of (*m/z* 727) (Schliemann et al., 2001). Furthermore, high-resolution LC-ESI-TOF-MS was used for completing the structure elucidation (Piattelli and Minale, 1964). Some cultivated species contained a higher percentage of acylated betacyanins than wild species, representing a potential new source of these pigments as natural colorants.

Celosianin and isocelosianin from *Celosia cristata* plumes (Lee et al., 1986) were found to be *p*-coumaroyl and feruloyl derivatives of I and II 5-*O*-glucuronosylglucosides. In addition to the known compounds amaranthine (Fig. 2.57) and betalamic acid, the structures of three other pigments were elucidated to be immonium conjugates of betalamic acid with dopamine, 3-methoxytyramine (I), and (S)-tryptophan by various spectroscopic techniques.

2.15.1 Ultraviolet & visible spectra of Celosia cristata

The colorant showed (*Celosia cristata*) one major peak, λ_{max} at 535 nm (betacyanins). The dye did not show much difference in the visible spectrum at pH 4 and 7; however, the peak shifted λ_{max} to 533 nm at pH 9 as showed in Fig. 2.58.

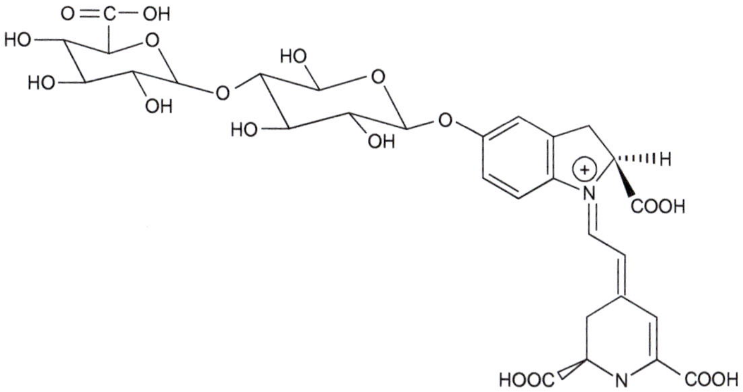

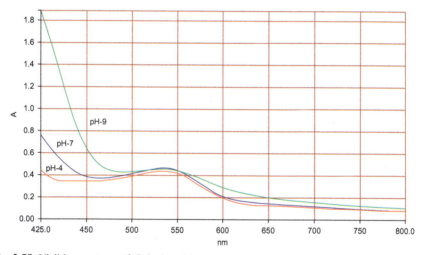

Fig. 2.57 Amaranthine.

Fig. 2.58 Visible spectrum of *Celosia cristata*.

2.16 Chemical characterization of Nerium

Four flavonoids (Paris and Duret, 1972) with rutin, quercetin, and quercetrin, among them, were detected, isolated, and identified in pink flowers of *Nerium*. From the petroleum ether extract of the pink flowers, β-sitosterol was isolated. The constituents detected were β-sitosterol, stigmasterol, campesterol, β-amyrin, ursolic acid, choline, and sucrose. Formation of acetate esters of the sterols enabled their separation (Figs. 2.59 and 2.60).

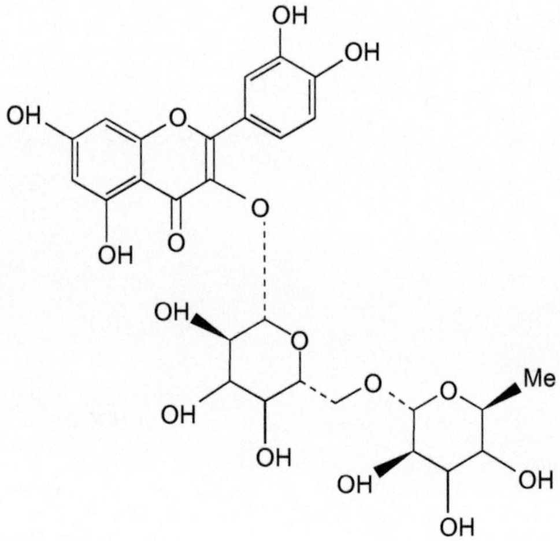

Fig. 2.59 Quercetin 3-*O*-rutinoside.

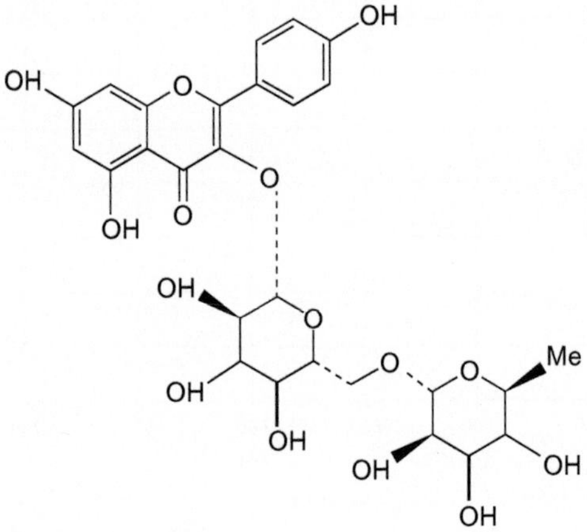

Fig. 2.60 Kaempferol 3-*O*-β-rutinoside.

2.16.1   *Ultraviolet & visible spectra of Nerium*

The colorant showed one major peak, λ_{max} at 321.43 nm in the UV region (flavonoids) and at 503.8 nm in the visible region (Fig. 2.61).

2.16.2   *HPLC of Nerium flower extract*

Solvent system: 95:5 MeOH-DW
  Run time: 15 min
  The sample was prepared in MeOH; chromatogram was taken at 255 nm and at 280 nm (Figs. 2.62 and 2.63).

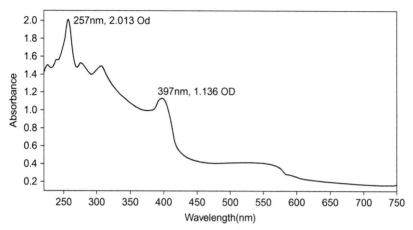

Fig. 2.61 UV-visible spectrum of *Nerium*.

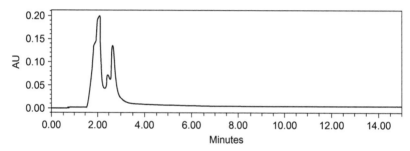

Fig. 2.62 HPLC of *Nerium* flower at 255 nm.

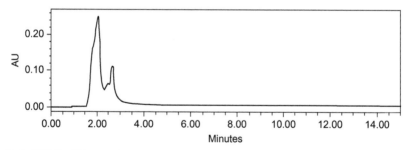

Fig. 2.63 HPLC of *Nerium* flower at 280 nm.

2.17 Chemical characterization of hollyhock

Petals of hollyhock (0.5 kg) were extracted twice with acetone-water (1:1) mixture (5 L) at 50–55°C for 20–25 min, the extract was filtered, and acetone was distilled off to give dry product (60 g) containing 12%–13% red pigment. Preliminary analysis showed (Salikhov and Idriskhodzhaev, 1978; Bhattacharya and Shah, 2000) that the pigment contained cyanidin-3-glucoside, delphinidin-3-glucoside, and malvidin-3,5-diglucoside as shown in Figs. 2.64–2.66.

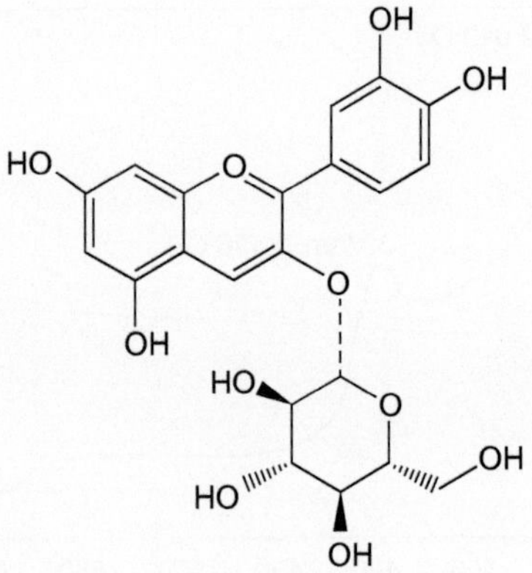

Fig. 2.64 Cyanidin-3-glucoside.

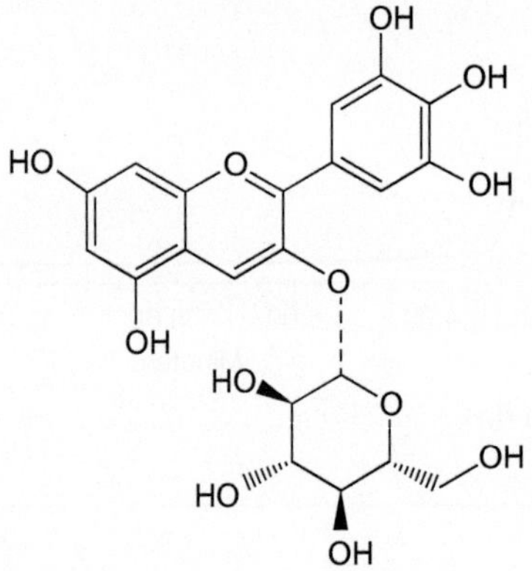

Fig. 2.65 Delphinidin-3-glucoside.

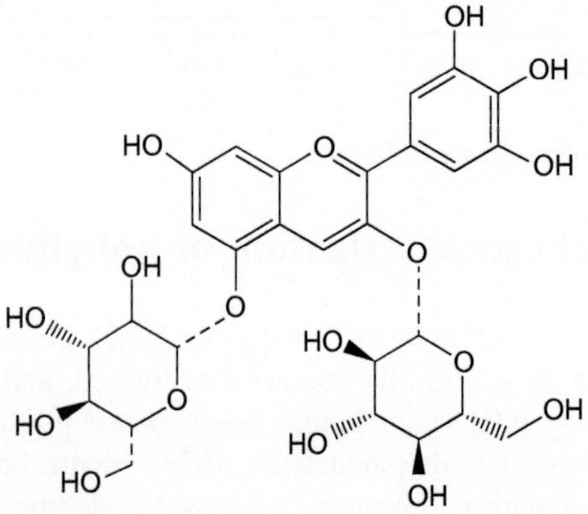

Fig. 2.66 Malvidin-3, 5-diglucoside.

2.17.1 Ultraviolet & visible spectra of hollyhock

See Figs. 2.67–2.69.

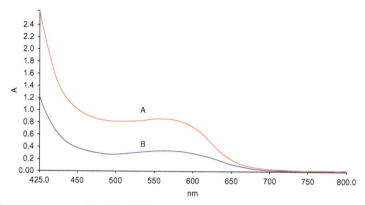

Fig. 2.67 Visible spectra of hollyhock flower extract.

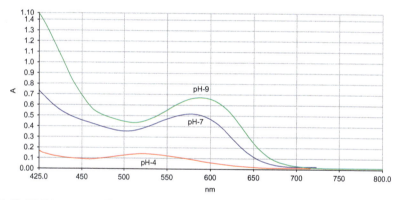

Fig. 2.68 Visible spectra of hollyhock at different pH.

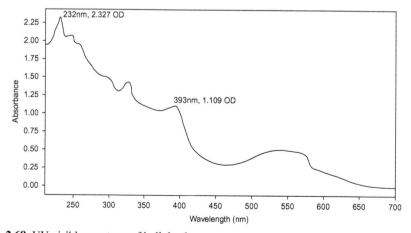

Fig. 2.69 UV-visible spectrum of hollyhock.

2.17.2   HPLC of hollyhock

Solvent system: 95:5 MeOH-H₂O
   Run time: 15 min
   The sample was prepared in MeOH; chromatogram was taken as real run (Fig. 2.70).

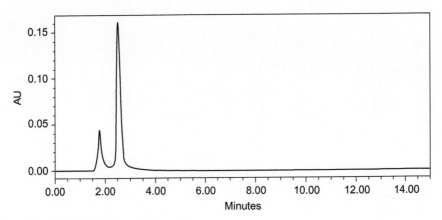

Fig. 2.70 HPLC of hollyhock flower extract.

2.18   Chemical characterization of *Hibiscus mutabilis* flower

Petals of *Hibiscus mutabilis* (0.5 kg) were extracted twice with acetone-water (1:1) mixture (5 L) at 50–55°C for 20–25 min, the extract was filtered, and acetone was distilled off to give dry product (60 g) containing 12%–13% red pigment. Preliminary analysis (Subramanian and Nadx, 1970; Ishikura, 1973; Yao et al., 2003) showed that the pigment contained 4-5 nonacosane, β-sitosterol, betulinic acid, hexyl stearate, stigmasta-3,7-dione, stigmasta-4-ene-3-one tetratriacontanol quercetin, and kaempferol. Isolation and purification were carried out on silica gel or polyamide column chromatography. The constituents were identified by physicochemical properties and spectral analysis. Ten compounds were obtained, nine of them were determined as tetracosanoic acid β-sitosterol, daucosterol, salicylic acid, emodin, rutin, kaempferol-3-*O*-β-rutinoside, kaempferol-3-*O*-β-robinobioside, and kaempferol-3-*O*-β-D-(6-E-*p*-hydroxycinnamoyl)-glucopyranoside (Figs. 2.71–2.73). All compounds were isolated from the plant for the first time except β-sitosterol and salicylic acid. The new flavonol glucoside quercetin 3-sambubioside 6 (I) and isoquercitrin, hyperin, guaijaverin, and a compound yielding kaempferol, glucose, galactose, and xylose on acid hydrolysis were isolated from the ethyl acetate extract of pink petals of *Hibiscus mutabilis* (Lim, 2014).

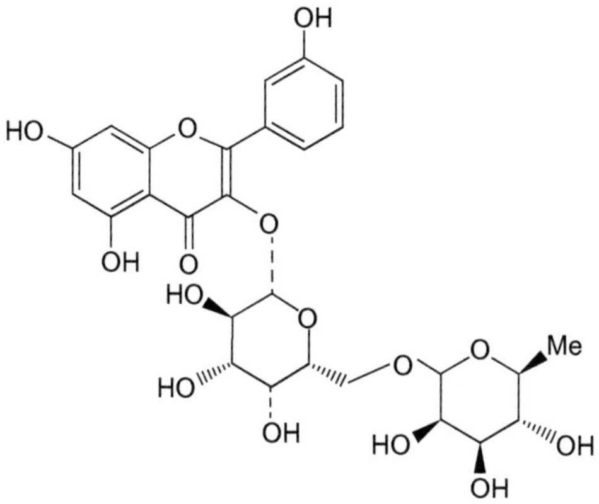

Fig. 2.71 Kaempferol-3-*O*-β-D-robinobioside.

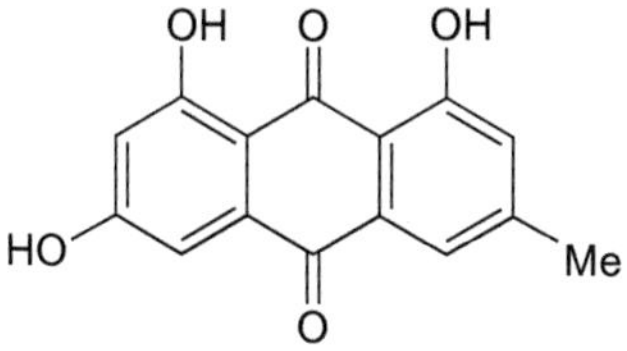

Fig. 2.72 Emodin.

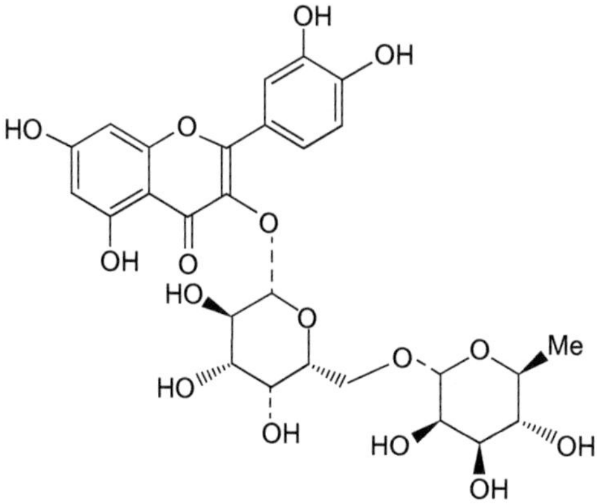

Fig. 2.73 Rutin.

2.18.1   *Ultraviolet & visible spectra of Hibiscus mutabilis flowers*

Plant source were crushed and dissolved in distilled water and allowed to boil in a beaker kept over water bath for quick extraction for 3 h. All the color was extracted from flowers by the end of 3 h. The solution was filtered for further use. The colorant showed one major peak, λ_{max} at 585 nm in the visible region. Comparison of the color content in dry and fresh flowers was also carried out as shown in graph I. The extract shows slight changes in different pH as shown in the Fig. 2.74. At pH 4, 7, and 9, the λ_{max} was at 585.64 nm; however, their absorbance were different—0.48, 0.67, and 0.96, respectively (Fig. 2.75).

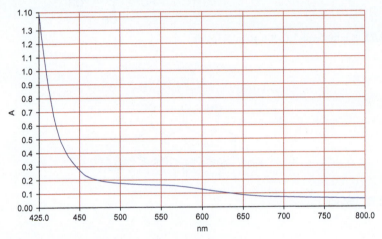

Fig. 2.74 Visible spectra of *Hibiscus mutabilis* flowers.

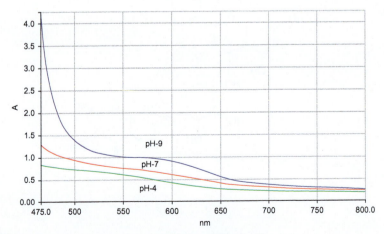

Fig. 2.75 Visible spectra of *Hibiscus mutabilis* at different pH.

2.19   Chemical characterization of *Cayratia carnosa* fruit extract

The blue to red color imparted by the anthocyanins depends largely upon the pH of the medium (Furia, 1977). The anthocyanins normally exist as glycosides; the aglycone component alone is extremely unstable. The anthocyanin pigments present in grape-skin extract consist of diglucosides, monoglucosides, acylated monoglucosides, and acylated diglucosides of peonidin, malvidin, cyanidin, petunidin, and delphinidin (Figs. 2.76 and 2.77). The amount of each compound varies depending upon the variety of wild blue berries and climatic conditions (Harborne, 1958; Ishikura and Sugahara, 1979; Timberlake and Bridle, 1975).

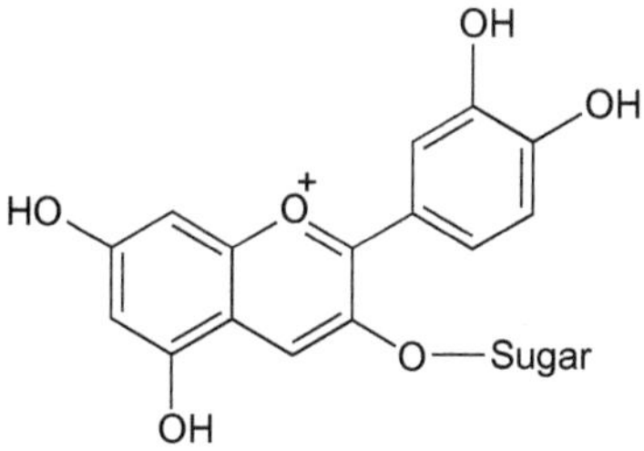

Fig. 2.76 Cyanidin (a).

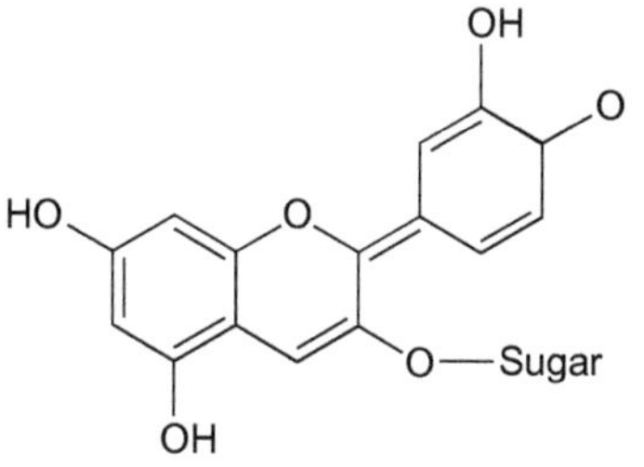

Fig. 2.77 Cyanidin (b).

2.19.1   *Ultraviolet & visible spectra of Cayratia carnosa*

Fruits of *Cayratia* plant source were crushed and dissolved in distilled water and allowed to boil in a beaker kept over water bath for quick extraction for 3 h. All the color was extracted from fruits by the end of 3 h. The extraction of anthocyanins from fruit skins is comparatively simple. The dark bluish-purple skins of the berries are separated from the rest of the fruit, freed completely from pulp by hydraulic pressure, and are extracted with water without delay and used for dyeing. The solution was filtered for further use. The colorant showed one major peak, λ_{max} at 585 nm as shown in Fig. 2.78. The dye did not show much difference in the visible spectrum at pH 4 and 7; however, the λ_{max} shifted to 600 nm at pH 9. This dye is inexpensive and abundantly available, and the method of application is very simple, producing no pollutants.

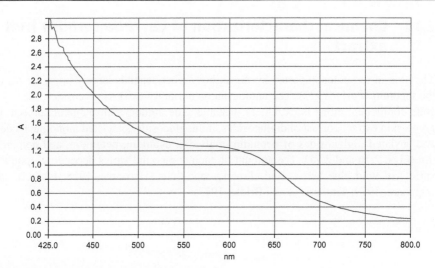

Fig. 2.78 Visible spectrum of *Cayratia carnosa.*

2.20   Chemical characterization of *Tegetus erecta* flower

Lutein was isolated according to the method developed (Francis and Markakis, 1989). Dry flowers were ground to 10–40 mesh size and dissolved in 8% ethanolic solution of KOH and washed with the same solution 4–6 times at 65–70°C under stirring. Addition of water followed by neutralization and filtration yielded an extract; this extract was further extracted with ethyl acetate 6–10 times; finally, crude lutein was obtained after distillation of ethyl acetate, and recrystallization from THF and deionized water yielded the final product of lutein crystal with a purity >90%. The isolated molecule was identified by NMR; mass spectra and the structure were shown in Fig. 2.79.

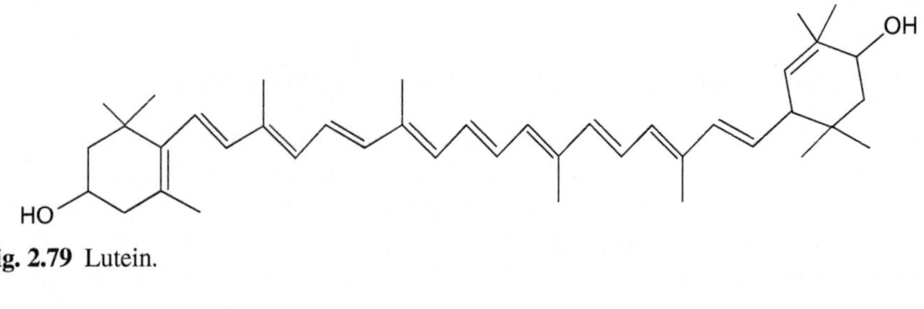

Fig. 2.79 Lutein.

Literature procedure of extraction of flavonoids from *Tagetes patula* L. (Li et al., 2007) showed preferential isolation of patulitrin and patuletin (Fig. 2.80) as main flavonoids and investigated their dyeing potential on wool using only alum mordant. The investigators had used a tedious method of macerating the flowers

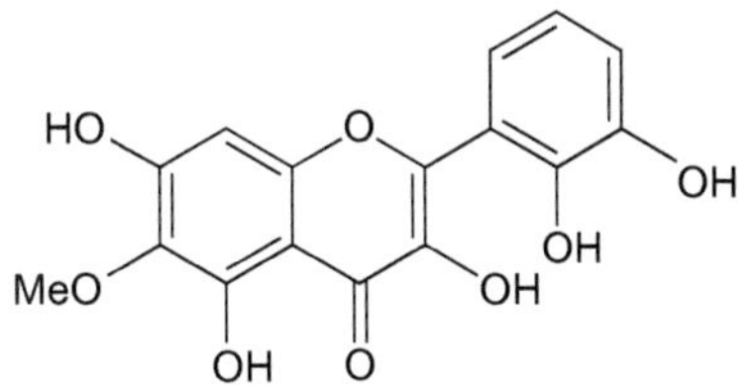

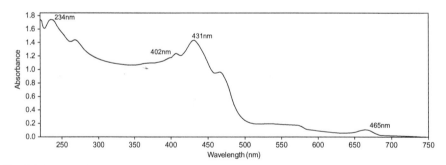

Fig. 2.80 Patuletin.

and had used solvents ethanol/water in varied composition 2:8, 3:7, and 5:5 v/v for extraction. However, we carried out column chromatography of crude extract on silica gel using the elution systems 20% EtOAc/hexane - 50% EtOAc/hexane and further on with 40% EtOAc from *Tagetes erecta* extract to isolate one major flavonoid—patuletin.

2.20.1 *Ultraviolet & visible spectra of* Tegetus

UV-visible spectrum of the yellow compound separated from *Tagetes* flower extract shows peak at 265, 353, and 392 nm (0.72, 0.50, and 0.55 A), respectively. All the peaks are characteristics of yellow flavonoid compounds (Fig. 2.81).

Fig. 2.81 UV-visible spectrum of *Tagetes*.

2.20.2 *HPLC of* Tagetes *extract*

Solvent system: 95:5 MeOH-DW
    Run time: 15 min
    The sample was prepared in MeOH; chromatogram was taken at 255 nm (a) and 280 nm (b) (Figs. 2.82 and 2.83).

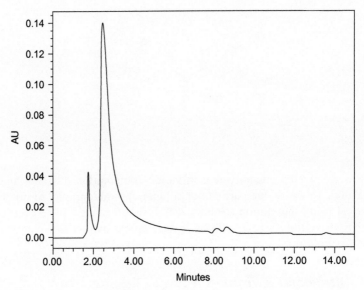

Fig. 2.82 HPLC of *Tagetes* at 255 nm.

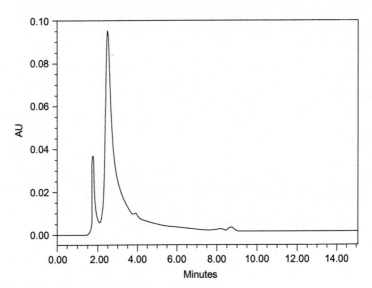

Fig. 2.83 HPLC of *Tagetes* at 280 nm.

2.21   Chemical characterization of Curcuma

Curcumin is the principal curcuminoid of the Indian curry spice turmeric, the other two curcuminoids being demethoxycurcumin and bis-demethoxycurcumin. The curcuminoids are polyphenols and are responsible for the yellow color of turmeric (Tang et al., 2002). Curcumin can exist in at least two tautomeric forms, keto and enol (Payton et al., 2007) as shown in Fig. 2.84.

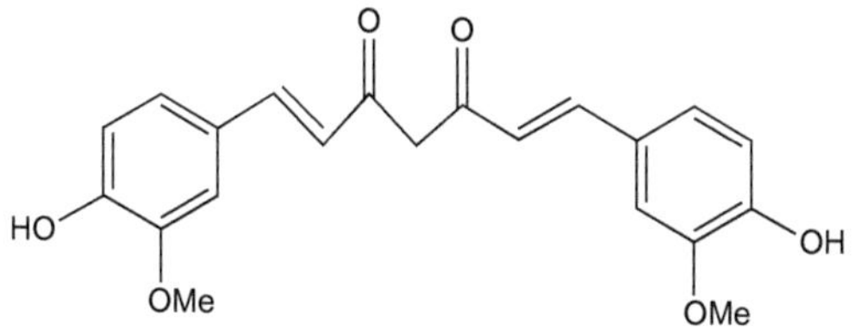

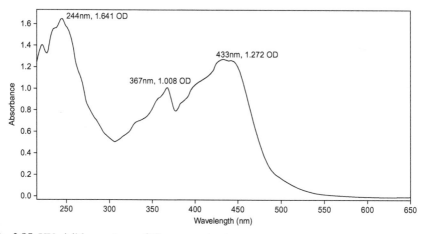

Fig. 2.84 Structure of curcumin.

2.21.1   Ultraviolet & visible spectra of Curcuma

See Fig. 2.85.

Fig. 2.85 UV-visible spectrum of *Curcuma* extract.

2.21.2   HPLC of Curcuma powder

Solvent system: 95:5 MeOH-H_2O

Run time: 15 min

The entire sample was prepared in MeOH, chromatogram taken at 255 nm. HPLC of curcumin was also studied by many researchers (Naidu et al., 2009) (Fig. 2.86).

The spectroscopic techniques are widely used in color identification of natural dye molecules. These are simple and reliable technique for measurement, quality control, and dynamic measurement, and they are easily available and accepted worldwide. Various natural dyes were identified and confirmed with these procedures.

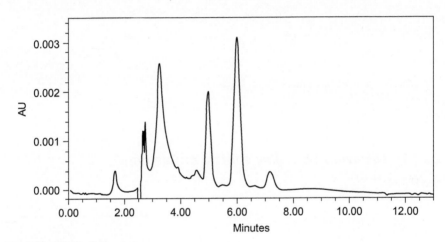

Fig. 2.86 HPLC of *Curcuma*.

References

Abrahart, E.N., 1977. Dyes and their intermediates. 62–63.

Bhattacharya, S.D., Shah, A.K., 2000. Metal ion effect in dyeing of wool fabric with Catechu. J. Soc. Dyers Chem. 116, 10.

Deo, H., Desai, B., 1999. Dyeing of cotton and jute with tea as a natural dye. Color. Technol. 115, 224–227.

Francis, F.J., Markakis, P.C., 1989. Food colorants: anthocyanins. Crit. Rev. Food Sci. Nutr. 28, 273–314.

Furia, T.E., 1977. Current Aspects of Food Colorants. CRC Press, Boca Raton, FL.

Gupta, R., Chandra, S., 1971. Chemical investigation of *Delonix regia* Raf. flowers. Indian J. Pharm 33 (4), 74–75.

Gupta, D., Dhiman, R., Naithani, S., Ahmed, B., 1988. Chemical investigation of *Bischofia javanica* Blume. Pharmazie 43, 222–223.

Harborne, J., 1958. Spectral methods of characterizing anthocyanins. Biochem. J. 70, 22.

Ishikura, N., 1973. Anthocyanins and flavonols in the flowers of *Hibiscus mutabilis* f. *versicolor*. Kumamoto J. Sci. Biol. 11, 51–59.

Ishikura, N., Sugahara, K., 1979. A survey of anthocyanins in fruits of some angiosperms, II. The Botanical Magazine = Shokubutsu-gaku-zasshi 92, 157–161.

Jung, S.J., Kim, D.-H., Hong, Y.-H., Lee, J.-H., Song, H.-N., Rho, Y.-D., et al., 2007. Flavonoids from the flower of Rhododendron yedoense var. Poukhanense and their antioxidant activities. Arch. Pharm. Res. 30, 146–150.

Jungalwala, F., Cama, H., 1962. Carotenoids in *Delonix regia* (Gul Mohr) flower. Biochem. J. 85, 1.

Kondo, T., Yoshikane, M., Goto, T., Yoshida, K., 1989. Structure of anthocyanins in scarlet, purple, and blue flowers of salvia. Tetrahedron Lett. 30, 6729–6732.

Latha, P., Panikkar, K., 1998. Cytotoxic and antitumour principles from *Ixora coccinea* flowers. Cancer Lett. 130, 197–202.

Latha, P., Nayar, M., Singh, O., George, V., Panikkar, K., Pushpangadan, P., 2001a. Isolation of antigenotoxic ursolic acid from *Ixora coccinea* flowers. Actual Biol. 23, 21–24.

Latha, P., Panikkar, K., Suja, S., Abraham, A., Rajasekharan, S., 2001b. Chemistry, pharmacognosy, pharmacology and botany of *Ixora coccinea*—A review. J. Med. Arom. Plant Sci 23, 670–676.

Lee, S.-Y., Shin, Y.-C., Byun, S.-M., Jo, J.-S., CHO, S.-J., 1986. Evaluation of red pigment of cockscomb flower in model food systems as a natural food colorant. Korean J. Food Sci. Technol. 18, 389–392.

Li, W., Gao, Y., Zhao, J., Wang, Q., 2007. Phenolic, flavonoid, and lutein ester content and antioxidant activity of 11 cultivars of Chinese marigold. J. Agric. Food Chem. 55, 8478–8484.

Lim, T., 2014. Hibiscus mutabilis. In: Edible Medicinal and Non Medicinal Plants. Springer International.

Montilla, E.C., Arzaba, M.R., Hillebrand, S., Winterhalter, P., 2011. Anthocyanin composition of black carrot (*Daucus carota* ssp. *sativus* var. *atrorubens* Alef.) cultivars antonina, beta sweet, deep purple, and purple haze. J. Agric. Food Chem. 59, 3385–3390.

Naidu, M.M., Shyamala, B., Manjunatha, J., Sulochanamma, G., Srinivas, P., 2009. Simple HPLC method for resolution of curcuminoids with antioxidant potential. J. Food Sci. 74, C312–C318.

Paris, R.R., Duret, S., 1972. Contribution à l'étude de la répartition et du métabolisme des flavonoïdes. Variations des flavonoides chez le Laurier-rose (*Nerium oleander* L.) au cours de la végétation. Bull. Soc. Bot. France 119, 531–542.

Payton, F., Sandusky, P., Alworth, W.L., 2007. NMR study of the solution structure of curcumin. J. Nat. Prod. 70, 143–146.

Piattelli, M., Minale, L., 1964. Pigments of centrospermae—II.: distribution of betacyanins. Phytochemistry 3, 547–557.

Purohit, A., Mallick, S., Nayak, A., Das, N., Nanda, B., Sahoo, S., 2007. Developing multiple natural dyes from flower parts of Gulmohur. Curr. Sci. 92, 1681–1682.

Ragasa, C.Y., Tiu, F., Rideout, J.A., 2004. New cycloartenol esters from *Ixora coccinea*. Nat. Prod. Res. 18, 319–323.

Salikhov, S., Idriskhodzhaev, U., 1978. Prospective coloring plant for the food industry. Khlebopek, Konditer, Prom-st. (8), 23–24.

Samanta, A.K., Agarwal, P., Singhee, D., Datta, S., 2009. Application of single and mixtures of red sandalwood and other natural dyes for dyeing of jute fabric: studies on colour parameters/colour fastness and compatibility. J. Text. Inst. 100, 565–587.

Schliemann, W., Cai, Y., Degenkolb, T., Schmidt, J., Corke, H., 2001. Betalains of *Celosia argentea*. Phytochemistry 58, 159–165.

Singh, R., Kalidhar, S., 2004. Chemical constituents of the stems of *Acacia arabica* (Lamk.). J. Indian Chem. Soc. 81, 436–437.

Subramanian, S.S., Nadx, A., 1970. A note on the colour change of the flowers of *Hibiscus mutabilis*. Curr. Sci. 39, 323–324.

Tanaka, T., Nonaka, G.-I., Nishioka, I., Kouno, I., Ho, F.-C., 1995. Bischofianin, a dimeric dehydroellagitannin from *Bischofia javanica*. Phytochemistry 38, 509–513.

Tang, B., Ma, L., Wang, H.-Y., Zhang, G.-Y., 2002. Study on the supramolecular interaction of curcumin and β-cyclodextrin by spectrophotometry and its analytical application. J. Agric. Food Chem. 50, 1355–1361.

Timberlake, C., Bridle, P., 1975. The anthocyanins. In: The Flavonoids. Springer, USA.

Tomás-Barberán, F.A., Harborne, J.B., Self, R., 1987. Dimalonated anthocyanins from the flowers of *Salvia splendens* and *S. coccinea*. Phytochemistry 26, 2759–2760.

Vankar, P.S., Shanker, R., Shalini, D., Mahantab, D., Tiwaric, S.C., 2007. Characterisation of the colorants from leaves of *Bischofia javanica*. Int. Dyer 192, 31–33.

Willstätter, R., Willstätter, R., Bolton, E., 1917. Untersuchungen über die Anthocyane. XI. Über das Anthocyan der rotblühenden Salviaarten. Justus Liebigs Annalen der Chemie 412, 113–136.

Yao, L.-Y., Lu, Y., Chen, Z.-N., 2003. Studies on chemical constituents of *Hibiscus mutabilis*. Chin. Tradit. Herb. Drug. 34, 201–202.

Structure-mordant interaction, replacement by biomordants and enzymes

P.S. Vankar
FEAT (Facility for Ecological and Analytical Testing), Kanpur Kalyanpur, India

3.1 Introduction

Today, in the world of growing environmental consciousness, natural colorants have attracted everyone's attention. It is only from the point of view of safety health and environment that natural dyes have gained momentum. Natural dyes are a class of colorants extracted from vegetative matter and animal residues. These are considered as mordant dyes as they require the inclusion of one or more metallic salts of aluminum, iron, chromium, copper, tin, and others for ensuring reasonable fastness of the color to sunlight and washing. These metallic salts combine with the dyestuff to produce dye aggregates, which cannot be removed from the cloth easily. These natural dyes are considered eco-friendly provided metallic salts used are the safer ones.

However, use of natural dye alone will not help the environment much because some of the mordants like chromium and copper are not so good for environment. Thus, using benign chemicals like aluminum, tin, and iron is the need of the hour. These chemicals help in the production of some pleasing colors without harming the eco-cycle at the least. Natural dyes require chemicals in the form of metal salts to produce an affinity between the cotton, silk fabrics, and wool yarn between the pigments, and these chemicals are known as *mordants*. Accurately weighed textile sample generally treated with 2%–4% of different metal salts such as stannic chloride, stannous chloride, alum, ferrous sulfate, copper sulfate, and potassium dichromate, prior to dyeing (Vankar, 2016). Different mordants give different colors with the same natural dye stuff. For example, cochineal, if mordanted with alum, will give a crimson color; with iron, purple; with tin, scarlet; and with chrome or copper, purple. The color imparted by dyestuff to the resulting solution depends on the electronic properties of the chromophore molecule. Dyestuff colorants not only tend to have excellent brilliance and color strength and are typically easy to process but also have poor durability, poor heat and solvent stability, and high migration (Gürses et al., 2016). Natural dyes are mostly nonsubstantive and must be applied on textiles by the help of mordants, usually a metallic salt, having an affinity for both the coloring matter and the fiber. Transition metal ions usually have strong coordinating power and/or capable of forming weak-to-medium attraction/interaction forces and thus can act as bridging material to create substantivity of natural dyes when a textile material being impregnated with such metallic salt (i.e., mordanted) is subjected to dyeing with different natural dyes, usually

Natural Dyes for Textiles. http://dx.doi.org/10.1016/B978-0-08-101274-1.00003-3

having some mordantable groups facilitating fixation of such dye (Samanta and Konar, 2011). Based on activity with mordants and fabric, dyes are classified based on both the structure of the dye and the way in which the dye is applied to the fabric:

- *Direct dyes*: They are charged, water-soluble organic compounds that bind to ionic and polar sites on fabric molecules. Direct dye molecules contain both positively and negatively charged groups and are easily absorbed by fabrics in aqueous solution. Simple salts such as sodium chloride and sodium sulfate may be added to the solution to increase the concentration of dye molecules on the fiber.
- *Substantive dyes*: They interact with fabrics primarily via hydrogen bonding between electron-donating nitrogen atoms (−N:) in the dye and polar−OH or−CONH−groups in the fabric.
- *Mordant dyes*: The ability of a dye to bond to a fabric may be improved by using an additive called a mordant. *Mordant dyes* are used in combination with salts of metal ions, typically aluminum, chromium, iron, and tin. The metal ions adhere to the fabric and serve as points of attachment for the dye molecules.

3.2 Mordanting

The first step of the actual dyeing process is mordanting. A mordant is a chemical that, when "cooked" with the fiber, attaches itself to the fiber molecules. The word "mordant" comes from the French "mordre," and mordants can be described as metallic salts with affinity for both fiber and dyes stuffs, and they improve the color fastness (Vankar, 2016). Dyes are categorized as either "mordant" or "adjective" or "indirect" dyes. Most of the natural dyes are mordant dyes except the very few direct dyes and vat dye such as indigo. The latter dye needs no mordants. The dye molecule, then, attaches itself to the mordant. Different mordants give different colors when combined with the same dye. For example, the dye, cochineal, when used with alum sulfate gives a fuchsia color; when used with tin, the color is more scarlet; and when used with copper, it is purplish. Mordants, except for alum and iron, are considered toxic and, therefore, should be avoided in the preparation of eco-textiles; otherwise, the whole exercise will be self-defeating. As the mordants are toxic to the dyer, the disposal of the bath becomes an environmental problem. Therefore, the choice of mordants is limited. Alum and iron are ideal safe mordants. Other chemicals known as assistants may be used in addition to dyes and mordants that help in the coloration of the fabric in one way or the other, for example, to change pH and hence the color, sometimes to brighten the color, to help in the absorption of the mordant metal, or to slow down the rate of absorption of pigments or for evenness. These include potassium hydrogen tartrate (cream of tartar), oxalic acid, tannic acid, acetic acid, formic acid, ammonia, sodium sulfate (Glauber's salts), sodium chloride (common salt), and sodium carbonate (washing soda). Treating cotton with tannic acid is useful as it prepares the fabric for effective absorption of the dye.

The effect of the mordant being distinctive is that it produced almost identical color strength on every fabric. Basically, natural dyes will not adhere to natural fibers without the use of a mordant or fixative. While you may initially get a beautiful result from the dyeing, it will soon wash out or fade away. Protein fibers like silk and wool absolutely need a plant extract (dye) and a mineral mordant. Some factors affect mordant working:

- Mordants should not affect the physical characteristics of the fibers.
- Sufficient time should be allowed for the mordant to thoroughly penetrate the fiber.
- If the mordant is only superficial, the dye will be uneven.

In addition to adding substances to a bath for mordanting, the vessel that is used may itself serve as mordant. This classical definition of mordant dyes has been extended to cover all those dyes that form a complex with metal mordant. The complex may be formed by first applying the mordant (premordanting), by simultaneous application of the mordant and the dye (simultaneous mordanting), or by after treatment of the dyed material with mordant (postmordanting).

3.2.1   General method of fabric mordanting

Accurately weighed fabric sample was treated with different metal salts, generally premordanting with metal salts was carried out before dyeing. The mordant dissolved in water to make liquor ratio 1:50. The wetted sample was then put into the mordant solution, and then, it was brought to heating. Temperature of this bath should be raised to 60°C over a period of 30 min and left at that temperature for another 30 min. The mordanted material was then squeezed and dried. Mordanted silk and wool need be used immediately for dyeing because some mordants are very sensitive to light.

3.2.1.1   Mordanting of cotton

Mordanting is very important for cotton dyeing. Cotton is generally pretreated with tannic acid salt to make it suitable for mordanting. Natural dyeing of cotton is more difficult than silk and wool. Cotton is not very porous and will not hold the dye stuff without a more complicated preparation for mordanting; the fiber must be cleaned first.

Preparation of alum mordant. To prepare alum mordant, first, alum powder and cream of tartar are mixed with little boiling water and then made up with the remaining required water. Stirring is continued well till the chemicals are dissolved and water should be lukewarm. The wetted fiber is put slowly to bring the bath to the required temperature. Dye the fiber immediately, and it should be dried in the shade and stored for further use.

Tin mordant. Dissolve cream of tartar or oxalic acid in a little quantity of hot water. When it is thoroughly dissolved, some more hot water is added. Addition of stannous chloride and mixing well are continued till it dissolves. The remaining water is added. When it is lukewarm thoroughly, loosened wetted fiber is put, and the heating to required temperature is done and continued to work for the specified time.

Copper mordant. Dissolving sulfate of copper in lukewarm water, and remaining required quantity of water is added. The fiber is entered and worked as tin mordant. If bright colors are desired, cream of tartar may be added in the beginning itself in the copper sulfate.

Chrome mordant. Mordanting with potassium dichromate is best just before dyeing. Dissolve the potassium dichromate in little warm water and make up the solution with the rest of required water. When the bath is lukewarm, then similar procedure is followed as stated above. Care should be taken while handling chrome mordant. Covering the pot with a lid except when the fiber is worked is necessary to avoid inhalation. Apart from this, chrome is very sensitive to light. If light falls on any part of the fiber, it will darken the fiber and result in uneven dyeing.

Iron mordant. Dissolve ferrous sulfate with a little warm and addition of cream of tartar to this, and this should be mixed well. Addition of the remaining total water entering the fiber and working for the specified time at specified temperature gives best results.

3.2.1.2 Mordanting in case of silk and wool

Accurately weighed silk and wool samples were mordanted as usual method. They are treated with different metal salts, before dyeing. The mordanted fabric was then squeezed and dried. Mordanted wool should be used immediately because some mordants are very sensitive to light.

3.2.2 Formation of metal complexes in dyeing

The bonding of dyes via metal complexes is an ancient technique known as mordanting (from French, *mordre*, to bite or catch). Otherwise, unreceptive cotton substrates can be activated by attaching a metal ion to the substrate before treating with the dye or by forming a metal-dye complex first and then attaching it to the substrate through the metal ion by further complex formation. Many metal ions (cations) form coordinate complexes with molecules that bear electron-donating groups, or anions. The complex-forming species are called ligands. and the electron-donating groups either have heteroatoms with nonbonding pairs of electrons (such as N and O) or carry a formal negative charge.

Metal complexes have very different chemical and physical properties from the parent metal ion. However, the effect on the absorption spectrum of the dye ligand is usually slight, resembling the changes seen on protonation, with only small changes in wavelength and absorptivity observed. The striking color changes observed on the formation of some complexes may be the result of charge-transfer transitions that involve the movement of electrons from the metal ion to the ligand; this occurs with some iron chelates.

3.2.3 Use of enzymes in natural dyeing

Enzymes have enjoyed considerable use for many years in the textile industry (Tsatsaroni and Liakopoulou-Kyriakides, 1995; Liakopoulou-Kyriakides et al., 1998); for example, amylases are used in desizing; cellulases are employed in denim finishing and the biopolishing of cellulosic fibers; proteases are used in leather, silk, and wool processing; and pectinases—amylase, lipase, and diasterase—are used in the biopreparation of cotton fabrics. We had used enzymes for dyeing cotton with two natural dyes, namely, *Tectona* and catechu (Vankar et al., 2007; Vankar and Shanker, 2008), and found very encouraging results. Wool was dyed with natural dyes such as juglone, lawsone, berberine, and quercetin using casein enzyme (Doğru et al., 2006). The results indicated that wool-lawsone-casein enzyme complex is the most appropriate one among all the tested enzymes on wool-dye complex as an insoluble substrate, emphasizing on the selectivity of the substrate dye with the enzyme.

3.2.4 Use of bio-mordant

Normally, metallic mordants such as Fe, Al, Cu, Pb, and Sn were used in dyeing. These metals might be found in plant although in small quantity, which act as minerals/nutrient for the plant to grow. They have a chelating agent characteristic that assist the process of dyeing. However, metallic mordant can cause serious bad effects to ecological (Rahman et al., 2013). In a newer approach, generally, plants having tannins, metal species, were used as biomordants. One of the important cases of such dyeing was using dye and biomordants of Arunachal Pradesh in India. A study to revive and restore the traditional dyeing practices of Arunachal Pradesh by using the traditional biomordant *Eurya acuminata* (Nausankhee or Turku) in place of metal mordant with *Rubia cordifolia* (Vankar et al., 2008). In another case, *Pyrus pashia* was used as a source of biomordant. Many *Pyrus* and *Prunus* species have been reported to contain copper such as the following: *Pyrus domestica* L. contains 0.33–34 ppm of copper, *Prunus serotina* or black cherry (stem) has 1.3–378 ppm of copper, *Prunus persica* (L.) or peach (fruit) has 0.3–30 ppm of copper; apart from this, other species are also known to contain Cu such as *Quercus phellos* L. or willow oak (stem), *Liquidambar styraciflua* L., *Brassica oleracea* L. var. *capitata* L., *Corylus avellana* L., and *Sassafras albidum* (http://www.levity.com/alchemy/metals_i.html). Biomordants have been used for more natural dyes (Guesmi et al., 2013). Many researchers have used biomordants in wool dyeing (Jayalakshmi and Amsamani, 2007; Jayalakshmi and Amsamani, 2008).

3.3 Use of metals, biomordant, and enzymes as mordant for different dyes

3.3.1 Metal complex with dye in Mahonia naupalensis

3.3.1.1 Nature of chelation in Mahonia naupalensis

Mahonia napaulensis contains mainly natural dye extracts that have less affinity for cotton fibers; their fastness was often enhanced by metal mordants, which form an insoluble complex with dye molecules, which include potassium aluminum sulfate (alum) and ferrous sulfate. The nature of the mordant-dye complex for *Mahonia* is shown below in Fig. 3.1A and B.

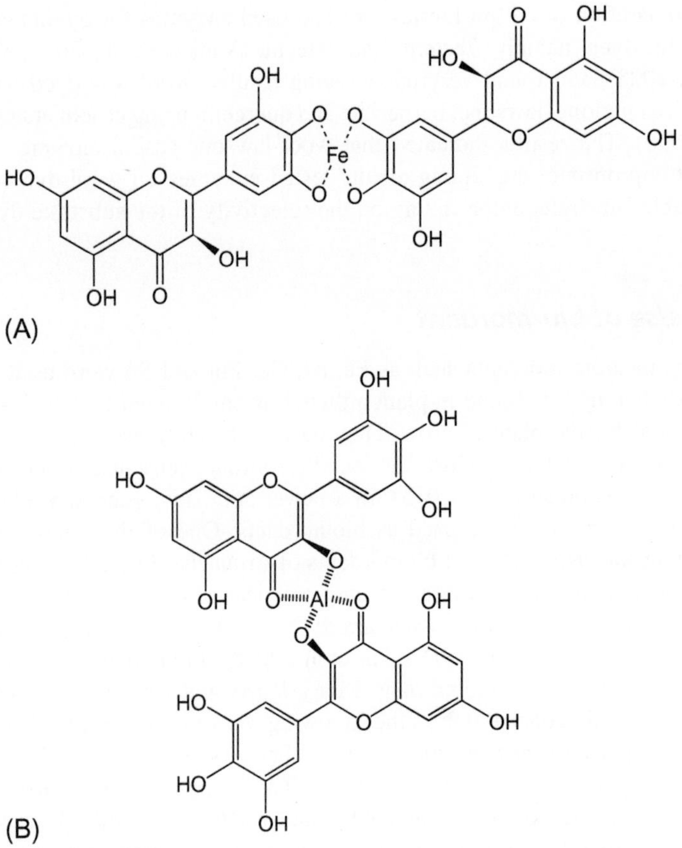

Fig. 3.1 (A) Mordant-*Mahonia* dye complex with iron (Fe). (B) Mordant-*Mahonia* dye complex with aluminum (Al).

3.3.2   Metal complex with dye in Rhododendron

3.3.2.1   Metal and anthocyanin conjugation in Rhododendron

Anthocyanin-metal complex constitutes a viable alternative for color stabilization, particularly if the metals involved do not imply a risk for the environmental pollution. One of the main characteristics of anthocyanins and anthocyanidins with o-di-hydroxyl groups in the B ring (Cy, Dp, and Pt) is their ability to form metal-anthocyanin complexes (Sati et al., 2003). Some studies about the color stability in plants suggest that the blue colors are due to a complexation between anthocyanin and some metals such as Al, Fe, Cu, and Sn (Muhan and Sibiao, 1991). Production of metal-anthocyanin complexes was suggested by changes in color of the samples as shown by L, a*, b*, and hue angle h values.

Possible role of Sn in chelation. The role of Sn metal with the *Rhododendron* colorant molecule can be envisaged as shown in Fig. 3.2. Role of Sn^{2+} on the thermal degradation of cyanidin 3-sophoroside was studied (Meng et al., 2005). It was found that anthocyanin pigments in strawberry, raspberry, and cherry preserves could be stabilized by the addition of alum and stannous and stannic chloride salts. The flavylium cation forms stable complex with Sn^{2+}, which in turns interact with the –OH groups of cellulose of cotton or $-NH_2$ groups of silk through oxochromes of the colorant. These are referred as inner complexes. Sn^{2+} forms a violet complex with the red flavylium cation. Some metals such as Fe^{3+} and Al^{3+} form stable deeply colored coordination complexes (Sadano, 2005) with anthocyanins that bear ortho-dihydroxyphenyl structure on the B ring as shown in Fig. 3.2.

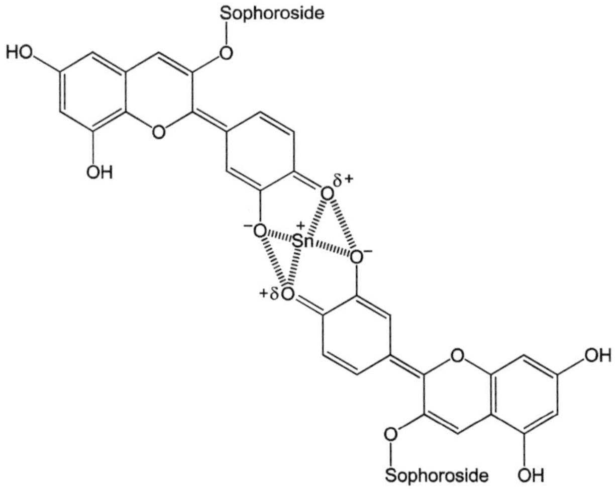

Fig. 3.2 Possible mode of chelation in *Rhododendron.*

3.3.3 *Metal complex with dye in* Hibiscus

Anthocyanin extract from *Hibiscus* flower formed a purple color complex with Sn^{2+} (Fig. 3.3).

Intramolecular effects such as copigmentation of anthocyanin-metal complexes play a vital role in the formation of the rich color. A narrow pH domain (pH 2–4) in which color amplification due to complexation is at a maximum has been found to give bright red color to the extract. Therefore, due to the enormous potential of natural anthocyanins as healthy pigments, there is increasing number of reports found in the literature on diverse fields such as development of analytic techniques for their purification and separation, applications in food (Cooper-Driver, 2001), identification and distribution in plants (Harborne and Williams, 2000; Harborne and Williams, 2001), quantitative analysis using chromatographic and electrophoresis techniques (da Costa et al., 1998; Shanker and Vankar, 2007) dyed wool with crude anthocyanin extract of hibiscus flowers.

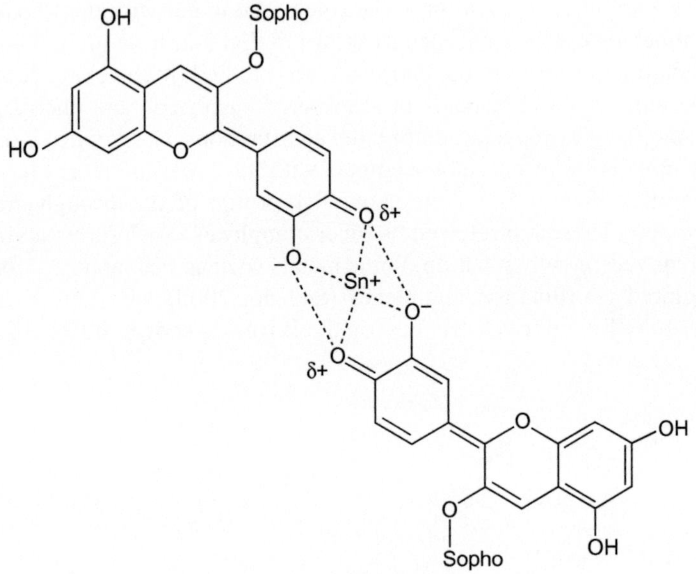

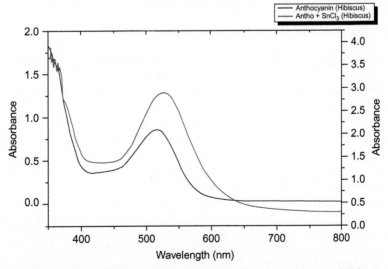

Fig. 3.3 Tin complex of cyanidin-3-sophoroside in *Hibiscus*.

3.3.3.1   Change in λ_{max} with addition of stannous salt

The intrinsic pH of the anthocyanin extract from hibiscus is 2.55 and shows an intense peak at λ_{max} 520 nm. Shifting of λ_{max} with change in pH is very prominent in the case of hibiscus flower extract. Stannous chloride lowered the pH of the anthocyanin extract from 2.55 to 1.81. At pH 3, 6, and 9, the λ_{max} shifts to 538 nm, 560 nm, and 552 nm, respectively, as shown Fig. 3.4 showed the change in visible spectra of *Hibiscus* anthocyanin

Fig. 3.4 Visible spectra of *Hibiscus* anthocyanin extract (left, *y* axis *black*) and *Hibiscus* anthocyanin-Sn extract (right, *y* axis *blue*).

extract and *Hibiscus* anthocyanin-Sn extract. Shift in λ_{max} by the addition of stannous salt is very apparent, which stabilizes the anthocyanin color on fabric in dyeing. The change in color profile after adding stannous to anthocyanin extract at different pH was due to change in λ_{max}, which was also responsible for good wash and light fastnesses.

3.3.4   Biomordant and enzyme complex with dye in Delonix regia

In this study, *Delonix* flower extract has been used as a source of natural dye for silk dyeing in conjunction with enzymes—protease, amylase, diasterase, and lipase, and the source of biomordant is *Pyrus pashia* (Vankar and Shanker, 2009; Vankar, 2013). The three main objectives using the extract of *Delonix* for the dyeing of silk fabrics are the following:

(i) To find an alternative to metal mordanting, making the dyeing process eco-friendly
(ii) To check the compatibility of enzymes and biomordant with *Delonix* dye as both enzyme and biomordant are known to be substrate specific
(iii) To carry out the dyeing process under mild operating conditions

3.3.4.1   Pyrus fruit chelation with dye

The high Cu content suggested stronger and useful chelation to the colorant for better dye adherence. The presence of the 4-oxo group in quercetin in conjunction with hydroxyl group also helps in chelation in flavonoids. Chelation of copper on the site between the 4-oxo group and C-5 OH group in flavonols and flavones has already been proposed (Mira et al., 2002; Brown et al., 1998). The number of OH groups is also important; the higher the number, the higher their chelating ability is. Thus, in the similar manner, the copper in *Pyrus* helps in chelation of Cu (II) with the colorant comprising mostly of flavones/flavonols. Based on the precedence, the probable mode of chelation of copper in *Pyrus* with quercetin has been proposed as shown in Fig. 3.5. Chemical binding of the dye with biomordant has also been proposed analogous to metal-dye complex formation.

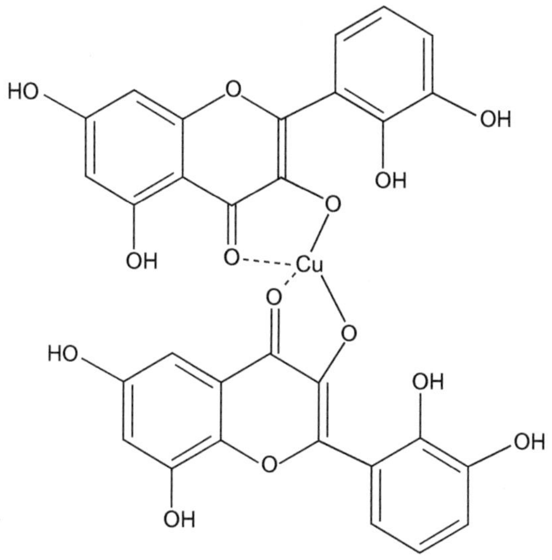

Fig. 3.5 Chelation of Cu in *Pyrus* with quercetin in *Delonix* and *Cayratia*.

3.3.5 Metal complex with the dye of Bischofia

Most of the natural dye extracts have poor affinity for cotton fibers; their fastness is often enhanced by metal mordants, which form an insoluble complex with dye molecules, as shown for potassium aluminum sulfate (alum) and ferrous sulfate in Figs. 3.6 and 3.7. The nature of the mordant-dye complex is shown below in Figs. 3.6 and 3.7:

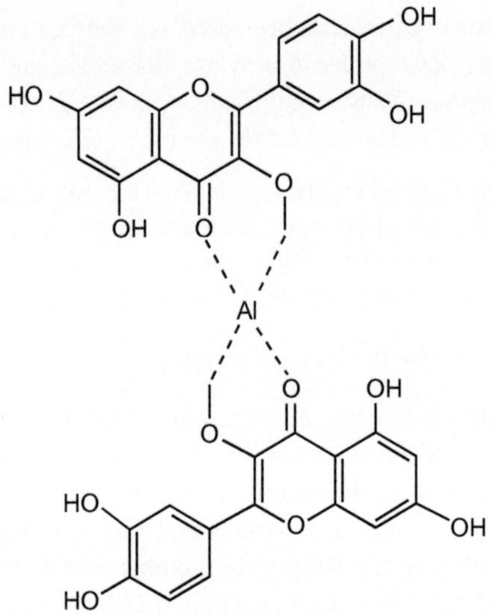

Fig. 3.6 Chelation with aluminum mordant in *Bischofia* (mordant-dye complex).

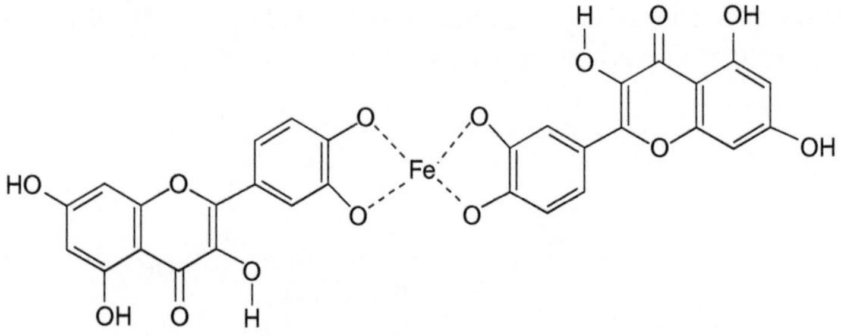

Fig. 3.7 Chelation with iron mordant in *Bischofia*.

The dye molecules are capable of forming six-member chelate rings with aluminum ions. The orthodihydroxy structure in the flavone dye molecules greatly enhances the chelation. The methanolic extract from *B. javanica* gave five fractions. Each of the color components isolated from leaves of *Bischofia javanica* showed a characteristic visible spectra for quercetin.

3.3.6 Metal complex of the dye of Rubia cordifolia

Rubia cordifolia contains mainly alizarin as well as purpurin, pseudopurpurin, munjistin, and rubiadin; because anthraquinone dyes have poor affinity for cotton fibers, their fastness was often enhanced by mordants. Mordants, which are metal salts that form an insoluble complex with dye molecules, include potassium aluminum sulfate (alum) and ferrous sulfate. The nature of the mordant-dye complex is well documented in the literature as shown in Fig. 3.8.

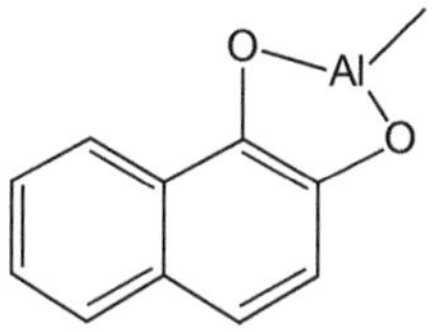

Fig. 3.8 Mordant-dye complex (chelation with aluminum mordant in *Rubia cordifolia*).

The alizarin molecules are capable of forming six-member chelate rings with aluminum ions. Colored lakes formed by the metal ions and dye molecules resist extraction by water and organic solvents, which readily strip similarly structured acid dyes. The sheer size of the complex may account for some of its insolubility. It is also likely that the large complexes are physically trapped within the fiber. The orthodihydroxy structure in the hydroxylanthraquinone molecules could greatly enhance the chelation. Atomic absorption spectroscopic analysis (GBC Avanta, model Sigma, Australia) of *Eurya acuminata* leaves extract showed 11.767 mg/L of Al content. The high Al content has been suggested to provide useful chelation to the anthraquinone moiety of *R. cordifolia* at two different sites, one with carbonyl and hydroxyl and the other with dihydroxyl moieties as shown in Figs. 3.9 and 3.10 (Vankar et al., 2008).

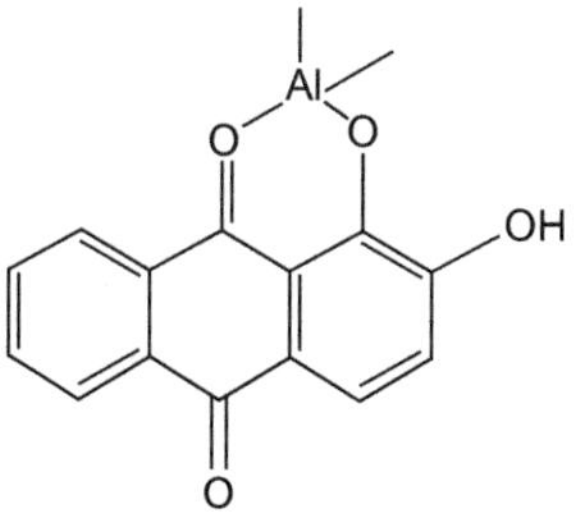

Fig. 3.9 Site of aluminum chelation in *Rubia cordifolia* (carbonyl and hydroxyl moieties).

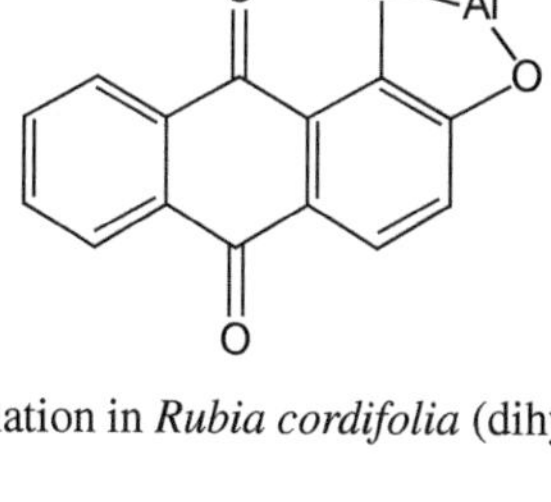

Fig. 3.10 Site of aluminum chelation in *Rubia cordifolia* (dihydroxyl moieties).

3.3.7 Metal complex of the dye of Tegetus erecta

The *Tagetus* dye extract also showed less affinity for cotton fibers as compared with wool and silk. The fastness was enhanced by using metal mordants, which formed an insoluble complex with dye molecules; a representative metal salt-M has been shown in Fig. 3.11. The stepwise dyeing of cotton fabric with tannic acid pretreatment, metal mordanting and dyeing with *Tagetus* flower extract, showed very good results of even dyeing. It has been proposed that the probable method of chelation of the dye molecules was occurring through the formation of bonds with metal ions. The dihydroxy flavone structure in one of the dye molecule patuletin has been shown to chelate with the metal mordant. This is a plausible explanation for chelation with one such flavonoid molecule. However, *Tagetus* flower extract should not be considered as a dye molecule; rather it should be attributed to a group of molecules that are cumulatively responsible for the dyeing process and color content (Lin et al., 2005; Vankar, 2009).

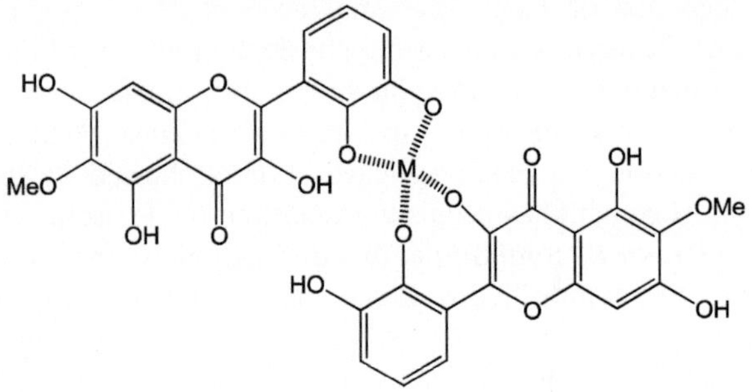

Fig. 3.11 Proposed dye-metal mordant complex in *Tagetus*.

3.4 Conclusion

Until the latter half of the 19th century, people were using natural dyes for coloring of textiles. Different parts of the plant were used to obtain various shades (Tesfaye et al., 2015). After invention of synthetic dyes, natural dyes are not used because of the advantage of synthetic dye over natural dye in respect to application, color range, fastness properties, and availability.

With the advent synthetic dyes, the baleful influence of the synthetic dyes affected the scope and use of natural dyes in middle of the 19th century. The one-time empathy between man and nature was marred due to synthetic dyes. The use of natural dyes has perished out, except in few pockets. The unbridled use of synthetic dye and the nontreatment of effluents contained in the waste waters of the dyeing process have lead to horrendous results and a terrible load on the environment. But there is increasing awareness among people toward natural products. Due to their nontoxic properties, low pollution, and less side effects, natural dyes are used in day-to-day food products as well (Siva, 2007; Zarkogianni et al., 2011). Demand of natural colors has been greatly increased in textiles, cosmetics, leather, food, and pharmaceutical industries; it is high time to switch over to natural colors. Earthy shade are sometimes called pastels but when combine with mordants, they change in to beautiful bright colors. Their colors are soothing to eyes, earthy warm, highly appealing, and noncarcinogenic, and above all, they perpetuate an ancient tradition and have potential colors of TOMORROW.

Further Reading

- Wilfred Ingamells, Colour for Textiles: A User's Handbook, The Society of Dyers and Colorists, 1993, ISBN 0 901956 56 2.
- John Shore (ed): Cellulosics Dyeing. The Society of Dyers and Colorists, 1995, ISBN 0 901956 68 6.
- Cristina Barrocas Dias, Marco Miranda, Ana Manhita, António Candeias, Teresa Ferreira, and Dora Teixeira 2013 Identification of Onion Dye Chromophores in the Dye Bath and Dyed Wool by HPLC-DAD: An Educational Approach, Journal of Chemical Education 90 (11), 1498-1500.
- P.S. Vankar, 2016. Handbook on Natural Dyes for Industrial Applications (Extraction of Dyestuff from Flowers, Leaves, Vegetables). NIIR Project Consultancy Services.

References

Brown, E.J., Khodr, H., Hider, C.R., Rice-Evans, C.A., 1998. Structural dependence of flavonoid interactions with Cu^{2+} ions: implications for their antioxidant properties. Biochem. J. 330, 1173–1178.

Cooper-Driver, G.A., 2001. Contributions of Jeffrey Harborne and co-workers to the study of anthocyanins. Phytochemistry 56, 229–236.

Da Costa, C.T., Nelson, B.C., Margolis, S.A., Horton, D., 1998. Separation of blackcurrant anthocyanins by capillary zone electrophoresis. J. Chromatogr. A 799, 321–327.

Doğru, M., Baysal, Z., Aytekin, Ç., 2006. Dyeing of wool fibres with natural dyes: effect of proteolytic enzymes. Prep. Biochem. Biotechnol. 36, 215–221.

Guesmi, A., Ladhari, N., Hamadi, N.B., Msaddek, M., Sakli, F., 2013. First application of chlorophyll-a as biomordant: sonicator dyeing of wool with betanin dye. J. Clean. Prod. 39, 97–104.

Gürses, A., Açıkyıldız, M., Güneş, K., Gürses, M.S., 2016. Dyes and pigments: their structure and properties. In: Dyes and Pigments. Springer International Publishing.

Harborne, J.B., Williams, C.A., 2000. Advances in flavonoid research since 1992. Phytochemistry 55, 481–504.

Harborne, J.B., Williams, C.A., 2001. Anthocyanins and other flavonoids. Nat. Prod. Rep. 18, 310–333.

Jayalakshmi, I., Amsamani, S., 2007. Bio-mordant for wool. Man-Made Text. India 50, 267–270.

Jayalakshmi, I., Amsamani, S., 2008. Dyeing wool using bio-mordants. Colourage 55, 102–105.

Liakopoulou-Kyriakides, M., Tsatsaroni, E., Laderos, P., Georgiadou, K., 1998. Dyeing of cotton and wool fibres with pigments from *Crocus sativus*—effect of enzymatic treatment. Dyes Pigments 36, 215–221.

Lin, M., Wu, D.-Q., Fan, Y., Feng, E., Yang, Y., 2005. Effects of microwave on extraction of *Tagetes erecta* flavonoid glycosides. J. Huzhou Teach. Coll. 2, 019.

Meng, H., Liu, G.-L., Zou, X.-Y., Jin, Q.-H., 2005. Microwave-assisted extraction of flavonoids from *Rhododendron dauricum* L. Fenxi Kexue Xuebao 21, 673–675.

Mira, L., Tereza Fernandez, M., Santos, M., Rocha, R., Helena Florêncio, M., Jennings, K.R., 2002. Interactions of flavonoids with iron and copper ions: a mechanism for their antioxidant activity. Free Radic. Res. 36, 1199–1208.

Muhan, Z., Sibiao, L., 1991. The effect of metal ions on mulberry and Rhododendron pigments. Shipin Yu Fajiao Gongye 6, 82.

Rahman, N.A., Tajuddin, R., Tumin, S., 2013. Optimization of natural dyeing using ultrasonic method and biomordant. Int. J. Chem. Eng. Appl. 4, 161.

Sadano, K., 2005. Health beverages containing rhododendron extracts, and method and container for their manufacture. Jpn. Kokai Tokkyo Koho 8, .

Samanta, A.K., Konar, A., 2011. Dyeing of Textiles With Natural Dyes. INTECH Open Access Publisher.

Sati, O., Rawat, U., Sati, S., Srivastav, B., 2003. Optimization of procedure for dyeing of wool with *Rhododendron arboreum* a source of natural dye. Colourage 50, 43–44.

Shanker, R., Vankar, P.S., 2007. Dyeing wool yarn with *Hibiscus rosa sinensis* (Gurhhal). Colourage 54, 66–69.

Siva, R., 2007. Status of natural dyes and dye-yielding plants in India. Curr. Sci. (Bangalore) 92, 916.

Tesfaye, T., Begam, R., Sithole, B.B., Shabaridharan, K., 2015. Dyeing cotton with dyes extracted from eucalyptus and mango trees. Int. J. Sci. Technol. 3, 310.

Tsatsaroni, E., Liakopoulou-Kyriakides, M., 1995. Effect of enzymatic treatment on the dyeing of cotton and wool fibres with natural dyes. Dyes Pigments 29, 203–209.

Vankar, P.S., 2009. Utilization of temple waste flower-*Tagetus erecta* for dyeing of cotton, wool and silk on industrial scale. J. Text. App. Technol. Manag. 6(1), 1–15.

Vankar, P.S., 2013. Handbook on Natural Dyes for Industrial Applications (with Colour Photographs). National Institute of Industrial Research, New Delhi.

Vankar, P.S., 2016. Handbook on Natural Dyes for Industrial Applications (Extraction of Dyestuff from Flowers, Leaves, Vegetables). NIIR Project Consultancy Services, India.

Vankar, P.S., Shanker, R., 2008. Ecofriendly ultrasonic natural dyeing of cotton fabric with enzyme pretreatments. Desalination 230, 62–69.

Vankar, P.S., Shanker, R., 2009. Potential of *Delonix regia* as new crop for natural dyes for silk dyeing. Color. Technol. 125, 155–160.

Vankar, P.S., Shanker, R., Verma, A., 2007. Enzymatic natural dyeing of cotton and silk fabrics without metal mordants. J. Clean. Prod. 15, 1441–1450.

Vankar, P.S., Shanker, R., Mahanta, D., Tiwari, S., 2008. Ecofriendly sonicator dyeing of cotton with *Rubia cordifolia* Linn. using biomordant. Dyes Pigments 76, 207–212.

Zarkogianni, M., Mikropoulou, E., Varella, E., Tsatsaroni, E., 2011. Color and fastness of natural dyes: revival of traditional dyeing techniques. Color. Technol. 127, 18–27.

Dyeing application of newer natural dyes on cotton silk and wool with fastness properties, CIE lab values and shade card

D. Shukla, P.S. Vankar
FEAT (Facility for Ecological and Analytical Testing), Kanpur Kalyanpur, India

4.1 Introduction

Color was considered by ancient people as a basic necessity as essential as food and water. The ancient people used exclusively dyestuffs of vegetable, mineral, and animal origin, all easily obtained in their own vicinity. Natural vegetable dyes have been used in most of the ancient civilizations in different countries, e.g., India, Egypt, Greece, and Rome. In India, use of vegetable dyes in dyeing, painting, and printing goes back to the prehistoric periods. The knowledge and use of color or dye on cotton, wool, and silk began with the dawn of the civilization and was first developed in the East, particularly in India. India has the long rich tradition of colored fabric design. There are many plants and some animal sources in nature that yield color and can dye fabric, leather, hair, and other items.

4.1.1 Advantages of natural colors/vegetable dyes

- Natural dyes bearing Ecomark are eco-friendly and acceptable in today's world.
- They are nontoxic and nonallergic and hazard-free for skin.
- Fastness can be achieved by the use of proper mordants.
- They save life, environment, fuel and time, and other investment process.

For successful introduction of vegetable dyes into technical dyeing processes, some additional demands have to be fulfilled:

- Increase of the number of available vegetable dyes with acceptable fastness properties suited for one—bath dyeing processes
- Formation of an efficient supplier organization that is able to provide a dye house with standardized dyes of constant quality and to generate an inventory of suitable vegetable dyes from application point of view
- Availability of technical information about the use of the dyes collected from the forests or locally grown plantation; emphasis be made on production of plant material in sufficient amounts with modern agricultural methods, which would include simple and environmentally clean extraction methods, suiting the requirement of a dye house
- Determination of eco-friendliness of the vegetable dyes for suitability for wearing dyed fabrics
- Determination of biodegradability of the waste generated after dye extraction from the plant sources

Natural Dyes for Textiles. http://dx.doi.org/10.1016/B978-0-08-101274-1.00004-5

It is of utmost importance to know the structure of the dyes depending on the dye structure, the mordant, and dye uptake that is expected. Pretreatments are a very important part of vegetable dyeing.

4.1.2 Natural dyeing principles

Application of natural dyes in today's scenario makes use of modern science and technology not only to revive the traditional technique but also to improve its rate of production, cost-effectiveness, and consistency in shades. It, therefore, requires some special measures to ensure evenness in dyeing. Many factors have to be accounted for when one works with natural dyes. They are as follows:

4.1.2.1 Nature of material to be dyed

Animal proteins like wool and silk dye best in acidic conditions and are weakened by alkaline. If an animal protein is dyed in alkaline conditions, it is best to end with a diluted vinegar rinse to restore a slightly acidic pH to the fibers before they dry. Plant materials like cotton and flax dye best in alkaline (basic) conditions and are weakened by acids. If cotton is dyed in acidic conditions, it is best to end with a weak washing soda bath to restore the fibers to slightly alkaline before they dry (Vankar, 2016).

4.1.2.2 Measurements of mordants and dyestuffs

Most dyeing procedures specify ingredients by weight rather than measure. Procedure will also specify the amount of fibers to be dyed, or the other ingredients will be expressed as a ratio to fiber weight. This is because the amount of water in the dyebath will not affect how strongly the fiber takes color but the amount of dyestuff in the dyebath does. So if 1 g of fiber has to be dyed with 1 g of dyestuff and then one wants to reproduce the same color on 5 more grams of fiber, the amount of dyestuff should be multiplied by five times as well. The water should always be enough to let the fibers move around freely; water quantity should be sufficient to dip the fabric/fiber properly.

4.1.2.3 Temperature

Different dyes work better at different temperatures. Most plant dyes benefit from being heated, but some (i.e., madder) change colors if allowed to boil. Sappan wood also has a tendency to change color when heated for prolonged hours. Some dyes work best at lower temperatures (safflower and woad/indigo).

4.1.2.4 Agitation

For getting even dye uptake, one should move the fibers around as much as possible in the dye pot. Unfortunately, when wool is heated and agitated, it tends to fall, so one must be very careful about how much one should move it around. For most wool, heating and cooling the dyebath slowly and being gentle while moving the fibers are necessary to avoid felting.

4.1.2.5   Natural dyes are unpredictable

There are so many factors involved in the dyeing process that reproducing a color exactly can be very difficult unless those parameters are followed strictly. Some reasons for disappointing results could be insufficient heat or too much heat, accidental iron or other metal contamination in the water, bad growing conditions for the dye plant, plant harvested at the wrong time of year, dyestuff allowed to dry out, dyestuff kept in humid conditions, dyestuff too old, and dye obtained from different plantation in terms of climate and soil conditions. The point here is to list some reasons for failure, which one would face if one does not get the expected color; the most experienced dyers in the world get accidental color sometimes. One can overdye and get the desired colors.

4.1.2.6   Wet fibers look darker

When trying to achieve a certain color, it has to be always remembered that the color when wet will always appear darker and may not look as bright when the fibers dry. Also, some color will bleed out after rinsing the fibers. Always dyeing to a darker shade in the dye pot than what is required. Lifting the fiber out of the dye pot to "air" is often good for the dyeing process to check the color.

4.1.2.7   Rinsing/fixing

Fibers/dyed textile should be rinsed after they have been dyed, and some dyes will still bleed for several washings afterward. It is advisable to add some washing soda to plant fibers or some vinegar to animal fibers to return them to their optimum pH in the last rinse.

4.1.3   Safety measures required in natural dyeing

Because dyeing substances and mordants can be poisonous, there are some important rules to follow when dyeing:
1. Dyeing should never be done in cooking vessels.
2. All measuring and stirring spoons, scales, thermometers, jars, etc. should be separately used for dyeing purpose.
3. The work area should be covered.
4. Wearing gloves to avoid contact with the skin is necessary.
5. Dye in a well-ventilated area or outdoors.
6. Rinsing fibers thoroughly after dyeing to remove all excess chemicals is essential.
7. Do not inhale steam from your dyebaths.
8. If you experience any itching, burning, rash, or other reaction, get away from the dye bath.

4.1.4   Standardization of natural dyeing

Natural dyestuffs produce offbeat, one-of-a-kind colors. No two dye lots are identical, each having subtle differences due to impurities peculiar to the particular plant material used. The problem of course with raw botanicals is that the numerous chemical ingredients that make up plants vary widely. The variations occur not only between plants of the same species but also from part to part of the same plant so that, for instance, in madder, the dye is contained in the roots, not the leaves. The type and quantity of chemicals present are affected by such things as soil, species, weather, and time of harvest as well as the part of the plant used. The manner in which they are stored and processed also has a profound effect. Color varies greatly with plants grown in different areas, due to mineral content of the soil and various other factors of growth. To produce fully standardized eco-friendly vegetable dyes from potential herbal/vegetable resources, evaluation of natural colors for eco-friendliness, standardization of processes for various textile substrates, and large-scale production and development of commercial natural dyes must be conducted, and there are certain specifications that need to be followed. They are the following:

(1) Color
(2) Appearance
(3) Optical density
(4) Water soluble matter
(5) pH of water extract
(6) Ash content
(7) Color component and its tinctorial value
(8) Total suspended solid content

4.1.5   Factors influencing natural dyeing of fabric

- *pH*: This is of critical importance in ionic bonding as the degree of ionization of the functional reactive groups on the substrate and also often on the dye molecule depend on the hydrogen ion concentration of the dyebath.
- *Ionic strength*: The addition of neutral inorganic salts such as sodium chloride to dye can alter the intensity of dyeing. Although the presence of inorganic ions might be expected to reduce staining intensity (competition for the ionic sites, the equilibrium being determined by mass action), in some instances, the increase in ionic strength results in an increase in dyeing. This latter phenomenon can be explained using a two-phase ion-exchange model that regards the dyebath and fabric as being two distinct phases.
- *Molecular size*: Another important factor affecting dyeing are the molecular size (molecular weight) of the dye, which will affect the rate of diffusion of the dye into the fabric and size of the chromophores (conjugated systems) with consequent large hydrophobic groups that tend to overstrain, and hence the lack of specificity.
- *Temperature*: Increasing temperature increases reaction rates and also the rate of diffusion of dye into fabric.

4.1.6   General methods of preparation for fabric before natural dyeing

4.1.6.1   Treatment of fabric

Every fabric, namely, cotton, silk, and wool, needs to be pretreated before dyeing as scouring, desizing, or degumming. Cotton fabrics have to be pretreated with tannic acid as cotton has very low affinity for natural dyes. The tannins play an important role in cotton dyeing and are largely used for preparing cotton so as to enable it to retain coloring matter permanently. The metal tannates present on the fabric form insoluble lakes with the natural dyes during dyeing. Similar to cotton silk is degumming before mordanting. Mordanting is a required pretreatment for all the fabric.

4.1.6.2 Mordanting

Natural dyes require chemical in the form of metal salts to produce an affinity between the cotton fabric and the pigments, and these chemicals are known as mordants. Mordants are commercially available, commonly in the form of salts from metals such as chrome, copper, tin, iron, and aluminum. Other types of mordants that are not metal mordants are tannins, cream of tartar, baking soda, and vinegar. The latter two serve to change the alkalinity and acidity, respectively, of the dye, another property that influences the final color. Accurately weighed fabric sample was treated with different metal salts. Mordanted cotton and silk fabrics should be used immediately because some mordants are very sensitive to light. There are three ways of mordanting. Mordants and dyes may be applied in three ways. They are as follows:

- Premordanting, where the mordant is applied first, followed by dyeing
- Postmordanting, where dyeing is done first and then mordanting is carried out
- Simultaneous mordanting, where mordant and dye are mixed together and applied

Metals have relatively low energy levels, so their incorporation into a delocalized system results in lowering of the overall energy. The absorbance of the hue and thus its color is related to this phenomenon. Most of the mordants that are used for natural dyeing are not severely toxic.

4.1.7 Dyeing

Dyeing is a sequential process. After removing the impurity of fabric, it was mordanted and then dyed with aqueous solution of dye plant.

After dyeing, several other parameters have to be checked to establish fastness rating/performance of dyed material. After fastness testing of dyed material, CIE Lab values have to be taken, and shade cards are prepared.

4.1.8 Fixing of dyed fabric

Dyed fabric was fixed in dye-fixer solution (2% solution of sodium chloride). The dyed fabrics were dipped for 30–45 min in dye-fixing solution that consists of saturated sodium chloride solution (2% w/w with respect to the fabric) and then rinsed thoroughly in tap water, before leaving to dry in open air. Several researches have been carried out (Cristea and Vilarem, 2006).

4.1.9 Fastness testing of dyed samples

The dyed samples were tested according to Indian standard methods ((BIS), 1982). The specific tests were color fastness for light *IS-2454-85*, color fastness to rubbing

IS-766-88, color fastness to washing *IS-687-79*, and color fastness to perspiration, *IS-971-83*. There are specified instruments to do dyed fabric analysis, which are following:

Xenoster: It can be used to test the light fastness of the dyed fabric.
Wash wheel: Thermolab model can be used to test the washing fastness of the dyed fabric.
Perspirometer: Sashmira model can be used for the testing of perspiration fastness of the dyed fabric.
Crock meter: Ravindra Engg model can be used for testing the rubbing fastness of the dyed fabric.

4.1.9.1   Color parameters/CIE Lab values of dyed samples

The CIE Lab color scale can be used on any entity whose color may be calculated. It is used widely in many industries. It offers a standard scale for evaluation of color values. A Lab color scheme is a color scheme with dimensions L for lightness and a and b for the color dimensions, based on nonlinearly XYZ coordinates. The terminology originates from the three dimensions of the Hunter color scheme which originated in 1948 as the first color scheme, which are L, a, and b. However, nowadays, Lab is usually an informal abbreviation for the L a*b* representation of the CIE 1976 color scheme (or CIE Lab) where the asterisks/stars are used to distinguish the CIE version from Hunter's original version. The difference between the original Hunter and CIE color coordinates is that the CIE coordinates are based on a cube root transformation of the color data, while the Hunter coordinates are based on a square-root transformation. Latest version of CIE Lab includes the CIE 1994 color scheme and the CIE 2000 color scheme.

The extension of L a*b* color scheme includes all perceivable colors, which means that its gamut exceeds those of the RGB and CMYK color schemes far more. One of the most important attributes of the L a*b* model is independent of the device used to ascertain it. The scheme itself is a three-dimensional real-number scheme the L, a*, and b* values are usually absolute, with a pre-defined range. The lightness, L, represents the darkest black at $L = 0$, and the brightest white at $L = 100$. The color channels, a* and b* represent true neutral gray values at $a* = 0$ and $b* = 0$. The red/green opponent colors are represented along the a* axis, with green at negative a* values and red at positive a* values. The yellow/blue opponent colors are represented along the b* axis, with blue at negative b* values and yellow at positive b* values. The scaling and limits of the a* and b* axes will depend on the specific implementation of Lab color.

CIE Lab scheme is relative to the white point of the XYZ data they were converted from. Lab values do not define absolute colors unless the white point is also specified. Often, in practice, the white point is assumed to follow a standard and is not explicitly stated (e.g., for "absolute colorimetric" rendering intent, the International Color Consortium L a*b* values are relative to CIE standard illuminant D50). The lightness correlate in CIE Lab is calculated using the cube root of the relative luminance (Koh and Hong, 2016).

Color matching system: The reflectance of dyed fabrics was measured on a Premier Colorscan.

4.1.9.2 Color measurements

The relative color strength of dyed fabrics expressed as K/S was measured by the light reflectance technique using the Kubelka-Munk equation (Kubelka, 1948; Kubelka, 1954). The reflectance of dyed fabrics was measured on a Premier Colorscan:

$$K / S = (1-R)^2 / 2R$$

where R is the decimal fraction of the reflectance of dyed fabric and K/S was measured. The CIE Lab values were determined for controlled, modified, and differently mordanted dyed fabrics.

The results show that the acacia bark is eco-friendly and can be used for commercial purpose to prepare value addition products in the handloom sector.

4.1.9.3 Dye exhaustion

The dye exhaustion percentage ($E\%$) was calculated according to the following equation:

$$\%E = [A0 - Ar / A0] \times 100$$

where $A0$ and Ar are, respectively, the absorbance of the dye bath before and after dyeing at λ_{max} of the dye used. The absorbance was measured on a spectrophotometer at λ_{max} of the dye used.

This is how the whole dye process can be carried out and can be standardized and various beautiful colors can be obtained.

References

(Bis), I.S.I., 1982. Handbook of Textile Testing. Manak Bhawan, New Delhi.

Cristea, D., Vilarem, G., 2006. Improving light fastness of natural dyes on cotton yarn. Dyes Pigments 70, 238–245.

Koh, E., Hong, K.H., 2016. Functional fabric treatment using tannic acid and extract from purple-fleshed sweet potato. Text. Res. J. 0040517516639829.

Kubelka, P., 1948. New contributions to the optics of intensely light-scattering materials. Part I. JOSA 38, 448–457.

Kubelka, P., 1954. New contributions to the optics of intensely light-scattering materials. Part II: nonhomogeneous layers. JOSA 44, 330–335.

Vankar, P.S., 2016. Handbook on Natural Dyes for Industrial Applications (Extraction of Dyestuff from Flowers, Leaves, Vegetables). NIIR Project Consultancy Services.

Dyeing of cotton by different new natural dyeing sources

D. Shukla, P.S. Vankar

FEAT (Facility for Ecological and Analytical Testing), Kanpur Kalyanpur, India

4A.1 Dyeing of Acacia bark on cotton

The prewashed cotton samples were mordant with $FeSO_4$, $SnCl_2$, $SnCl_4$, and alum. They were then dyed by aqueous extract of Acacia bark powder (% dye) (Tiwari et al., 2003). Dyeing was carried out in a sonicator for one hour and showed good dye uptake. The dye bath was reused in some cases or more dye extract was added to the previous bath so that maximum dye uptake is facilitated.

For attempting repeatability of shades in dyeing it is important to follow the given procedure.

- **i.** Only stainless steel dye bath should be used.
- **ii.** Water hardness should not be more than 300 ppm.
- **iii.** Fabric/Yarn water ratio should be 1:20.

4A.1.1 Fastness properties of dyed fabrics

It was observed that dyeing with Acacia bark powder gave fair to good fastness properties in sonicator dyeing. The color range with Acacia bark varies from skin color in the case of stannous chloride to dark brown in the case of ferrous sulfate. The color fastness to washing was between 4-5 as presented in Table 4A.1. Apart from this, the light fastness, rubbing, and perspiration fastness were also above average. A marked improvement in these properties is observed in sonicator dyeing of cotton fabric. Overall, it could be said that, Acacia bark dye can be considered for commercial purpose, the dyed fabric attains acceptable range.

Soaking and extraction of natural dye in sonicator on large scale is easy that not only results in low investment but also a reduction in natural contaminants. Such an extraction provides the advantage of not requiring further purification of the dyes and using the dye at low cost. Dyeing with Acacia bark powder in sonicator can be industrially very economical and useful. One sonicator machine can be used for dual purpose-for extraction as well as dyeing.

Natural Dyes for Textiles. http://dx.doi.org/10.1016/B978-0-08-101274-1.00009-4

Table 4A.1 **Fastness properties of Acacia dyed cotton**

Mordant used	Dyed shade	Wash	Light	Fastness rubbing		Properties perspiration	
				Dry	Wet	Alkaline	
		IS-687-79	IS-2454-85	IS-766-88		IS-971-83	Acidic
Stannic chloride	Light brown	4–5	4–5	4–5	3–4	4/4	4/4
Alum	Mud	4–5	4–5	4–5	4	4–5/4–5	4/4–5
Ferrous sulfate	Brown	4	4	3–4	3–4	4–5	4/4–5
Stannous chloride	Skin	4–5	4–5	4	4	3–4	3–4

4A.2 Dyeing of *Mahonia* on cotton

4A.2.1 *Extraction of the dye*

Mahonia stem pieces were completely dried at room temperature. The dried stems were crushed and ground to make powder. Dry stems from plant source were crushed and boiled in distilled water and heated to (100°C) in a beaker kept over water bath for quick extraction, for 1.5 h. All the color was extracted from leaves by the end of 1.5 h.

4A.2.2 *Special treatment of fabric before dyeing*

After removing the impurity of cotton fabric, it was treated with 4% (owf) sodium lauryl sulfate. Similarly washed cotton was treated with 1% Cellulase and another swatch was treated with ammonium sulfamate (4%).

4A.2.3 *Dyeing*

The pretreated fabric was used for dyeing with *M. naupalensis* extract (10%, owf). The dyeing time was 3 h at temperature 30–40°C in the case of conventional dyeing and 1 h in the case of sonicator. The dyed fabrics were dipped in dye-fix, which consists of saturated sodium chloride solution (2%, 15 min), and then rinsed thoroughly with tap water and dried in open air.

4A.2.3.1 *Effect of temperature*

The effect of temperature on the dyeability of the pretreated cotton fabrics with *Mahonia* dye was studied under Sonicator Dyeing (SD) and Conventional Dyeing (CD) conditions at different temperatures (30–40°C). The color strength increases with increase in dyeing temperature in both cases of SD and CD methods with pronounced increase in the SD case than the CD. This is due to the increase in dye-uptake that can be explained by fiber swelling and hence, enhanced dye diffusion. Also, the sonication provides other added advantages of no aggregation formation of dye molecules and thus leading to further enhancement of dye diffusion and better dyeability than CD.

4A.2.3.2 Effect of dyeing time

The effect of dyeing time was studied at high concentration of the dye, i.e., 10 g/100 mL water to reveal the effect of sonication on the de-aggregation of dye molecules in the dye bath as indicated by higher dye-uptake. The color strength obtained increased as the time increased in both SD and CD methods with much higher color strength at all points in the SD case. In the case of CD, the decline in color strength started after 180 min. The decline in the dye ability may be attributed to desorption of the dye molecules as a consequence of long dyeing time.

The stepwise dyeing of cotton fabrics with alum metal mordant by the natural dye *Mahonia naupalensis*, showed very good results. The dye uptake in the three cases of pretreatment (SLS, Cellulase, and Ammonium sulfamate) was 70%, 60%, and 58% under sonication for cotton, while 45%, 38%, and 37% under conventional dyeing. With Sodium lauryl sulfate (anionic agent) the dye extract showed bathochromic shift due to the cationic nature of the dye. However cellulase and Ammonium sulfamate made marginal difference by pretreatment.

4A.2.4 Optimization of extraction condition

The optimization of aqueous extract prepared from dried *Mahonia* stem was followed as per the method described (Vankar et al., 2008a).

4A.2.5 Color measurements and CIE Lab values

4A.2.5.1 Effect of mordanting conditions

It was observed that the pretreatment with three different types of chemicals imparted good fastness properties to the cotton fabric. Therefore, pretreatment of the cotton fabric with ammonium sulfamate, sodium lauryl sulfate, and cellulase enzyme showed the following sequence: SLS > Cellulase > Ammonium sulfamate in cotton for *Mahonia naupalensis*, the absorption of color by cotton fabric was enhanced by using these pretreatments, this might be due to the better penetration of the dye, maximum absorption, and easy formation of complexes with the fabric.

4A.2.5.2 Fastness testing

The dyed samples were tested according to Indian standard methods (Table 4A.2).

Mahonia naupalensis was found to have good agronomic potential as a dye crop in Arunachal Pradesh. Metal mordant when used in conjunction with pretreatment with SLS, Cellulase, or Ammonium sulfamate with *Mahonia naupalensis* stem extract was found to enhance the dyeability and fastness properties. Enhancement of dye uptake under sonication was 25%, 23%, and 21% with alum mordant for cotton, respectively.

Table 4A.2 **Comparative fastness properties of differently modified dyed cotton, under conventional heating and sonication conditions of *Mahonia naupalensis***

Dyeing methods	Wash–perspiration–rubbing–light					
	WF	**Per$_{acidic}$**	**Per$_{basic}$**	**Rub$_{dry}$**	**Rub$_{wet}$**	**LF**
Conventional (controlled)	3	2–3	2–3	2–3	2–3	2–3
Sonication	4	4	4	3–4	3–4	3
Conventional (cellulase)	3–4	3	3	3–4	3–4	3–4
Sonication	4	4–5	4–5	4–5	4–5	4–5
Conventional (amm. sulfamate)	4	4	3–4	3–4	3–4	4
Sonication	5	5	5	4–5	4–5	5
Conventional (SLS)	3	3–4	3–4	3–4	3	3
Sonication	4	4	4	4	4	4

WF=wash fastness, Per $_{acidic}$=perspiration fastness under acidic condition, Per $_{basic}$=perspiration fastness under basic condition, Rub $_{dry}$=rubbing fastness in dry, Rub $_{wet}$=rubbing fastness in wet, LF=light fastness.

The color range in cotton fabrics is light yellowish brown to khaki brown, for silk it ranges from light orange to dark orange, similar colors were obtained for wool yarn as well. Even the fastness properties in this case showed good results. The stepwise process of dyeing with premordanting developed was for the ease of industrial application.

4A.3   Dyeing cotton with Salvia flowers

Flowers from plant source were crushed and dissolved in distilled water and allowed to boil in a beaker kept over water bath for quick extraction for 3 h. Conventional dyeing of cotton fabric with the aqueous extract of *Salvia* flowers (red variety) by using tin mordant was carried out. CIE Lab values, K/S values, and fastness properties of the dyed fabrics were ascertained. The dye showed promising results and acceptability for commercial dyeing (Vankar and Kushwaha, 2011). Tin mordant was chosen particularly as it gave stability to the extract and the color content deepened.

4A.3.1   Dyeing

Dyeing was done by conventional dyeing method. The cotton was dyed with dye extract, keeping M:L ratio as 1:30 however for cotton dyeing it was used directly and the pH was maintained at 4 by adding buffer solution (sodium acetate and acetic acid) as followed (Vankar and Srivastava, 2010). The dye extract was prepared by adding 20 g of the dried flower powder in 100 mL water (M:L::1:40). The dyed fabric was washed with cold water and dried at room temperature; it was then dipped in brine for dye fixing. The color strength was determined colorimetrically using Premier Colorscan at the maximum wavelength of the natural colorant.

4A.3.2 Fastness properties

It was observed that dyeing with Salvia flower extract gave fair to good fastness properties in conventional dyeing. The Tables 4A.3 and 4A.4 shows L, a*, and b* values and it can be seen that in case of tin mordant cotton lower values of L show darker shades while higher L values signify lighter shades. Similarly negative a* and negative b* represents green and blue respectively for cotton. The results clearly show that dyeing with the extract of Salvia flower extract shows better dye uptake, without the mordant but the dye did not adhere properly.

This process showed reduced dyeing time and being cost effective, as the flower is available in abundance during the flowering time. Overall, it could be used for commercial purposes; the dyed fabrics attain acceptable range in terms of fastness properties as shown through K/S values.

Sn (IV) also indicates that the coloristic efficiency or tinctorial value of this natural dye for cotton fabric is not high for these mordants despite the tannic acid pretreatment to cotton fabrics.

Table 4A.3 shows the colorimetric values of dyed cotton fabric with *Salvia* after pretreatment with tin mordant, the dyeing difference without and with mordant imparted a shade change from pure red to purplish red. The aqueous extract of *Salvia* flower showed higher hue angle for dyed and premordanted fabrics.

Table 4A.4 shows the colorimetric values of dyed silk fabric with *Salvia* after pretreatment with tin mordant, the dyeing difference without and with mordant imparted

Table 4A.3 L, a*, b* values for dyed cotton with Salvia

Mordant (premordanting)	Color obtained	L*	a*	b*	C	H	K/S
Fabric	White	83.44	−1.70	4.13	4.47	112.43	2.22
Dyed Fabric without mordant	Red	82.74	21.83	3.05	22.04	7.96	100.13
Dyed fabric with mordant Stannic chloride	Purplish red	81.74	19.76	−0.38	19.96	358.89	65.49

Table 4A.4 L*, a*, and b* values for dyed silk with Salvia

Mordant (premordanting)	Color obtained	L*	a*	b*	C	H	K/S
Fabric	Cream	76.63	−1.59	5.52	5.74	106.11	4.70
Dyed fabric without mordant	Red	76.63	27.40	5.56	27.96	11.48	102.03
Dyed fabric with mordant stannic chloride	Purplish red	74.73	24.30	−1.40	24.34	356.69	68.66

Table 4A.5 **Fastness properties of dyed cotton with Sn mordant**

Dyeing methods	Wash–perspiration–rubbing–light					
	WF^a	Per_{acidic}	Per_{basic}	Rub_{dry}	Rub_{wet}	LF^b
Cotton (control)	2–3	2	2	2–3	2–3	2
Cotton (SnCl₄)	4	4	4	4	4	4

^{a} Wash fastness.
^{b} Light fastness.

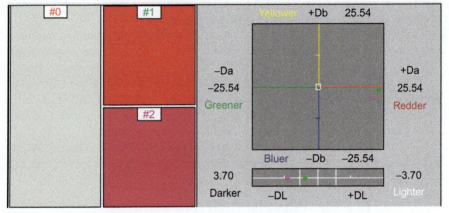

Fig. 4A.1 Difference in color obtained by tin mordant on cotton.

a shade change from pure red to purplish red. The aqueous extract of *Salvia* flower showed higher hue angle for dyed and premordanted fabrics (Table 4A.5).

Fig. 4A.1 shows the colorimetric values of dyed cotton fabric with *Salvia* after pretreatment with tin metal mordant (#0-2 in the order of Fabric, unmordanted, dye, and stannic chloride). The dyeing with Sn imparted a shade change from red to purplish red.

Aqueous extract of Salvia flowers yield brown to green shades on cotton with good fastness properties. The color strengths (K/S values) are good particularly cotton mordanted by ferrous sulfate. The dye has good scope in the commercial dyeing of cotton fabric for garment industry. Thus Salvia can be popularized as a cheap source of natural dye.

4A.4   Dyeing of cotton fabric by Canna

After removing the impurity of cotton fabric it was then treated with 4% (owf) solution of tannic acid in water. The fabric should be dipped in tannic acid solution for at least 4-5 h. It is squeezed and dried. Premordanting was used for this study; fabric which was already treated with tannic acid was dipped in mordant (2% for alum and 1% for stannous chloride, stannic chloride, ferrous sulfate, copper sulfate, and potassium dichromate separately) solution and was kept at 30°C for one hour. It was squeezed and air dried (Ghorpade et al., 2000).

Canna dye (Srivastava et al., 2008) is acidic in nature and is generally converted to its sodium/ potassium salts, which are completely soluble in cold water and on the other hand pure dye is poorly soluble in boiling water. In order to improve the dissolution of *canna* dye it was dissolved as o/w microemulsion. Canna dye has no affinity for cotton as it does not possess any cationic sites for attachment. Thus it can be used as mordant dye with prior treatment to fabric.

4A.4.1 Dyeing

4A.4.1.1 Effect of dye bath pH

A considerable effect on the dyeability of cotton fabrics depends on the pH values of the dye bath have while using the *canna* dye extract. The effect of the dye bath pH can be attributed to the correlation between dye structure and cotton fabric. Since the dye used is sparingly soluble in water, containing OH groups, it would interact ionically with terminal carboxylic groups of tannic acid pretreated cotton fabrics. The anion of the dye has complex characters, and thus it is bound on the fabric. Table 4A.6 shows the pH of the extracts of canna dye in different medium and their absorbance at 358 nm.

4A.4.1.2 Effect of temperature

The effect of temperature on the dyeability of cotton fabrics with *canna* dye extract was determined at different temperatures (30–40°C). It was observed that the color strength increased with increase of dyeing temperature and reached a maximum value at 40°C.

4A.4.1.3 Effect of dyeing time

As shown in Fig. 4A.2 the color strength obtained was increased as time increases for up to 6 h, it then decreases; dyeing for 6 h gave high color strength values (Table 4A.7).

When fresh canna flowers were used for aqueous extract, the aqueous extract with oil/water emulsion was compared with ethanolic extract, the K/S showed decrease in the aqueous extract oil/water emulsion but marked difference was seen in K/S value of

Table 4A.6 **pH of extraction medium**

Medium	pH	λ_{max} 542 nm Absorbance	λ_{max} 358 nm Absorbance
Aqueous	5.87	0.268	0.828
Ethanolic	7.57	0.286	2.500
Oil/water	7.27	1.286	1.700

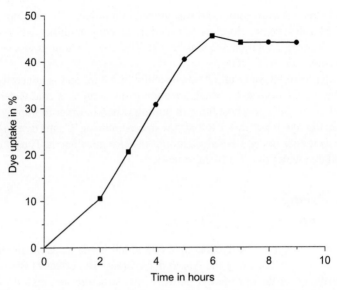

Fig. 4A.2 Effect of dyeing time on the color strength of dyed cotton fabrics.

Table 4A.7 L, a, and b values of alum mordanted cotton dyed fabric from different extracts

	L	a*	b*	C	H	K/S	δE*
Aq. ext (0)	59.463	1.326	20.560	20.603	86.275	25.47	–
Fresh o/w (1)	60.623	1.803	24.114	24.181	85.689	18.26	3.76
Ethanolic (2)	53.581	4.369	0.057	4.369	0.747	37.88	21.54

ethanolic extract as shown in Fig. 4A.3. Thus it was very clear that preparation of ethanolic extract, evaporation of ethanol to dryness and subsequent addition of water for dyeing showed best results. Even the color of the dyed fabric changed from greenish brown to purple in the latter case (Fig. 4A.4).

Result of ethanolic extraction of dye of *canna* flowers from fresh, frozen aqueous, fresh flower extract with oil/water microemulsion and frozen flower extract o/w microemulsion is shown in Table 4A.8. Best results were obtained with frozen extract and its oil/water microemulsion, even the dE* value is the highest in the case of frozen o/w. The same can be seen in the color obtained after dyeing as shown in Figs. 4A.5 and 4A.6.

From the Table 4A.8 it can be seen that the color strength and K/S has shown improvement, aqueous extract of *canna* has poor dyeability while ethanolic extract shows better results. But use of ethanolic extract is not viable in dye industry. Thus o/w emulsion offers the best results for use of canna as dye source. This is also shown in Figs. 4A.6 and 4A.7 and shades were shown in Fig. 4A.8.

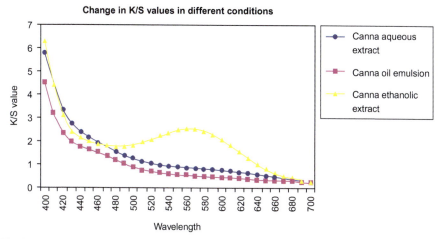

Fig. 4A.3 Change in K/S values of different extracts- aqueous, aqueous o/w and ethanolic.

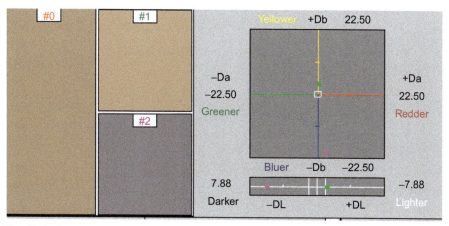

Fig. 4A.4 Colorimetric changes in fresh extracts—aqueous, aqueous o/w emulsion, ethanolic.

Table 4A.8 **L, a, and b values of alum mordanted cotton dyed fabric from different extracts**

	L	a*	b*	C	H	K/S	δE*
Fresh ext(0)	50.757	5.58	4.29	7.04	37.52	32.22	–
Frozen ext(1)	51.348	9.41	3.75	10.13	21.74	122.65	3.82
Fresh o/w (2)	50.559	5.08	6.47	8.24	51.90	30.51	2.32
Frozen o/w (3)	48.949	14.63	−3.65	15.07	345.98	181.24	12.17

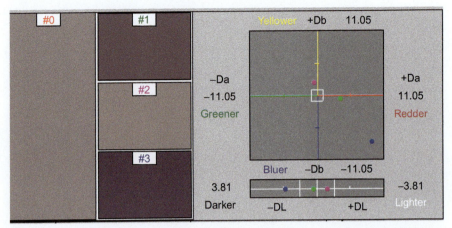

Fig. 4A.5 Colorimetric changes in fresh and frozen extracts with and without o/w emulsion.

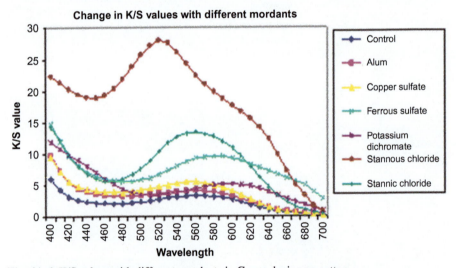

Fig. 4A.6 K/S values with different mordants in Canna dyeing on cotton.

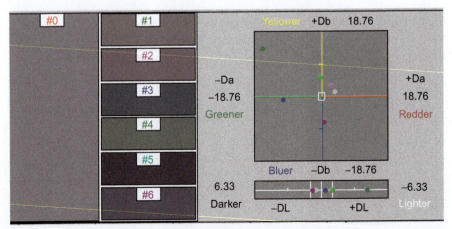

Fig. 4A.7 Colorimetric changes with different mordants with ethanolic extract/aqueous dyeing.

Control

Alum

Ferrous sulfate

Fig. 4A.8 Shade card of cotton dyed *Canna*.

Flavonoids and polyphenol compounds which were present in the crude canna extract could be used for dyeing cotton to get good fastness properties and high dye uptake. The highest dissolution and better dye uptake was observed for ethanolic extract of *canna* flowers, followed by O/W microemulsions which showed that for partially or completely insoluble dyes O/W microemulsions is a good choice for aqueous medium dyeing of cotton fabric. The conventionally dyed fabric with only aqueous medium showed uneven dyed surface, while the ethanolic and o/w microemulsion showed good dyeing results and color depth. Natural flavonoid dyes can provide bright hues and color fastness properties as shown in the dyed samples. They can serve as a noteworthy source of raw material in the future. Chemical modification of cotton fabric and innovative extraction of such natural compounds could be an interesting field of study as it could appreciably facilitate the use of such cheap and abundantly available dye source (Tables 4A.9 and 4A.10).

Table 4A.9 L*, a*, b*, C, and H values for dyed cotton with ethanolic extract of *Canna*

Method	Mordant	L*	a*	b*	C	H	Color %	K/S
	Controlled (0)	44.211	7.48	–3.04	8.08	337.86	100.00	45.36
	Alum (1)	45.293	7.12	2.49	7.55	19.30	141.06	63.99
Pre-treatment	Copper sulfate(2)	44.683	10.30	0.01	10.30	0.04	169.18	76.75
	Ferrous sulfate(3)	44.599	–3.45	–3.99	5.27	229.12	304.10	137.97
	Pot. dichromate(4)	48.545	–9.27	11.09	14.46	129.91	214.65	97.38
	Stannous chloride(5)	43.652	11.32	–1.79	11.46	350.98	781.55	354.57
	Stannic chloride(6)	43.344	8.24	–10.38	13.26	308.45	354.24	160.71

Table 4A.10 Effect of mordant on the visible spectrum of ethanolic extract of *Canna*

Absorbance	λ_{max} 352 nm	λ_{max} 542 nm	Shifted λ_{max}
Extract only	3.983	1.709	–
With alum	3.767	1.922	471 nm (0.212)
With ferrous sulfate	4.276	2.744	–
With copper Sulfate	3.636	1.962	–
With stannic chloride	3.804	3.876	555 nm (0.195)

4A.5 Dyeing with *Rhododendron*

Dark red variety of Rhododendron flowers were collected from the forests of Tawang, Arunachal Pradesh and were used for cotton dyeing.

4A.5.1 Dyeing

Dyeing was done by conventional dyeing method. The cotton was dyed with dye extract, keeping M:L ratio as 1:30; however for cotton dyeing it was used directly as followed in other work (Vankar and Shanker, 2009). The dye extract was prepared by adding 20 g of the dried flower powder in 100 mL water. The dyed fabric was washed with cold water and dried at room temperature; it was then dipped in brine for dye fixing. The color strength was determined colorimetrically using Premier Colorscan at the maximum wavelength of the natural colorant.

4A.5.1.1 Preparation and optimization of aqueous extract of Rhododendron

The flowers of Rhododendron were found to give out color in hot water very easily. The flowers were frozen after collection and then dipped in hot boiling water to get the maximum color in 30 min, which shows deepening of hue color. Increasing the quantity of flowers from 2 to 20 g per 100 mL water boiled for 60 min is accompanied with increase in color strength and depth in color hue.

4A.5.1.2 Optimization of mordants with K/S and Color hue changes

Premordanting the fabrics with different metal salts, i.e., $FeSO_4$, $SnCl_2$, $CuSO_4$, $SnCl_4$, $K_2Cr_2O_7$, and alum was carried out and then dyed by aqueous extract of *Rhododendron* to obtain different hue colors. Different mordants were used in 2%–4%, keeping in mind the toxicity factor of some mordants. As shown in the Fig. 4A.9 the different mordants not only cause difference in hue color and significant changes in K/S values but also L values and brightness index values.

4A.5.1.3 Fastness properties of dyed cotton

It was observed that dyeing with *Rhododendron* flower extract gave fair to good fastness properties in conventional dyeing. Dyeing for just one hour showed good dye uptake.

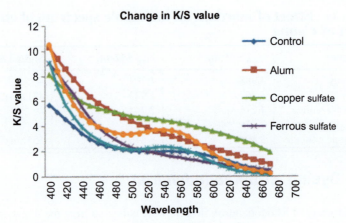

Fig. 4A.9 K/S values for dyed cotton fabric with *Rhododendron*.

Table 4A.11 **L*, a*, and b* values for dyed cotton with Rhododendron**

Mordant (premordanting)	Color obtained	L*	a*	b*	C	H	K/S
Alum	Olive green	51.36	5.61	28.35	28.90	78.76	145.75
Copper sulfate	Greenish brown	54.03	8.27	33.48	34.49	76.08	147.81
Ferrous sulfate	Khaki color	43.84	3.54	7.22	8.05	63.83	264.53
Potassium dichromate	Greenish Yellow	54.33	4.80	32.56	32.91	81.57	69.57
Stannous chloride	Olive green	56.16	0.77	31.13	31.14	88.53	59.84
Stannic chloride	Military green	58.35	−0.01	36.64	36.64	90.05	63.01

The Table 4A.11 shows L, a*, and b* values for cotton. The results clearly show that dyeing with the extract of Rhododendron flower extract shows better dye uptake, in just one hour. This process showed reduced dyeing time and cost effectiveness, as the flower is available in abundance during the flowering time. Overall, it could be used for commercial purposes; the dyed fabrics attain acceptable range in terms of fastness properties as shown through K/S values in Fig. 4A.9 for cotton. It was observed that dyeing with *Rhododendron* gave good fastness properties in conventional dyeing. The highest K/S for the ferrous sulfate in the case of cotton were obtained. The CIE Lab values for cotton samples are shown in Table 4A.11. Analysis of the data of Table 4A.11 reveals that the results of CIE Lab values and K/S values are consistent. The modest values of L in the case of Cr(VI), Sn(II), and Sn(IV) also indicate that the coloristic efficiency or tinctorial value of this natural dye for cotton fabric is not high for these mordants despite the tannic acid pretreatment to cotton fabrics.

Table 4A.12   Fastness properties of dyed cotton with different metal mordants

Dyeing methods	Wash–perspiration–rubbing–light					
	WF[a]	Per$_{acidic}$	Per$_{basic}$	Rub$_{dry}$	Rub$_{wet}$	LF[b]
Cotton (control)	2–3	2	2	2–3	2–3	2
Cotton (alum)	4	4	3–4	3–4	3–4	4
Cotton (FeSO$_4$)	4–5	4	4	4	4	4–5
Cotton (CuSO$_4$)	4	4	4	4	4	4

[a] Wash fastness.
[b] Light fastness.

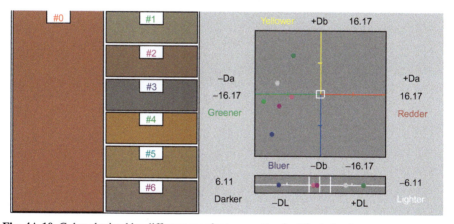

Fig. 4A.10 Color obtained by different mordants on cotton by *Rhododendron*.

Table 4A.12 shows the results of fastness properties determined for unmordated dye with some exemplary metal mordanted cotton and silk dyed fabrics.

Fig. 4A.10 shows the colorimetric values of dyed cotton fabric with *Rhododendron* after pretreatment with different metal mordants (#0-6 in the order of unmordanted, alum, copper sulfate, ferrous sulfate, potassium dichromate, stannous chloride and stannic chloride). The dyeing with different mordants imparted a shade range from olive green to khaki to light brown to military green having blackish tinge. The lightness value decreased for chromium, stannic, and stannous mordanted cotton sample and shade of depth retained their brightness, while the highest was obtained with ferrous sulfate having dullness and Fig. 4A.11 present shades obtained by *Rhododendron*.

Aqueous extract of *Rhododendron* flowers yield brown to green shades on cotton fabrics with good fastness properties. The color strengths (K/S values) are good particularly cotton mordanted by ferrous sulfate. The dye has good scope in the commercial dyeing of cotton for garment industry. This dye can be popularized as a cheap source of natural dye (Vankar and Shanker, 2010).

	Cotton
Control	
Alum	
Copper sulfate	
Ferrous sulfate	
Potassium dichromate	
Stannous chloride	
Stannic chloride	

Fig. 4A.11 Shade card of cotton dyed by *Rhododendron*.

4A.6   Dyeing cotton with cosmos

4A.6.1   Optimization of dyeing parameters

1. Optimum concentration of Cosmos dye. The concentration of dye material (Cosmos flowers) was optimized by taking two sets of four dye solutions, prepared by soaking and heating at 60°C. 2, 4, 6, and 10 g of sun dried cosmos flowers in 100 mL of water for cotton dyeing in microwave and sonicator. 10 g flowers in 100 mL water gave the best result. This optimization is with respect to dye uptake by the fabric and the shade required. For deeper shades larger amounts of flowers can be used.

2. Optimum time for extraction of Cosmos dye: Experimentation with 4 sets of sun dried Cosmos flowers placed in 100 mL of water and heated at 60°C in beakers. These beakers were removed from hot plate after 30 min, 40 min, 50 min, and one hour. It was observed that the optical density was maximum for the sample left for 1 h.

3. Optimum time and concentration for mordants: 2% solution of ecofriendly mordants such as alum, ferrous sulfate, stannous chloride, and stannic chloride were prepared. The cotton samples were first dipped in 4% tannic acid solution and then treated with metal salts for premordanting or postmordanting.

4.

    a) Optimum condition for microwave dyeing: The temperature was set at 80°C for 10 min (≈120 W) in the microwave. The cotton sample was dipped in dye solution and irradiated for 10 min. The dyed sample was rinsed and dried in shade.

    b) Optimum condition for sonicator dyeing: Cotton fabric is dipped in sonicator bath for one hour at room temperature. The dyed sample was rinsed and dried in shade.

4A.6.2   Dyeing with cosmos

The optimum concentration of the dye materials, time for dye extraction, mordanting, and dyeing by microwave and sonicator techniques have been studied. It has been observed that among the aqueous extraction method and methanolic extraction, the dye content is more in the methanolic extract. The dye uptake in microwave is better for methanolic extract dyeing as shown in Fig. 4A.12.

The Fig. 4A.12 shows that premordanting with stannic chloride is good however in Fig. 4A.13 postmordanting with stannous chloride show better fastness properties.

In Fig. 4A.14 comparative evaluation of fastness properties towards cotton fabric dyed by microwave and sonicator has been shown. Light fastness in case of microwave is 4 while for sonicator it is 3, washing fastness for the former is 4.5 and for the latter it is 4.25. Rubbing fastness both wet and dry are 4 and 3, respectively. Even fastness in perspiration in alkaline and acidic medium show similar differences. Microwave dyeing is better with regard to fastness properties.

The comparative fastness properties of cosmos in different premordanted cotton fabrics using sonicator is showed in Fig. 4A.14. Light fastness is best for stannic chloride,

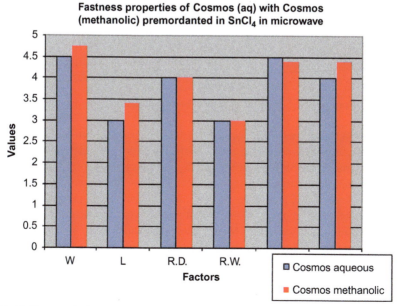

Fig. 4A.12 Dye uptake in microwave dyeing with cosmos (pre-mordanting).

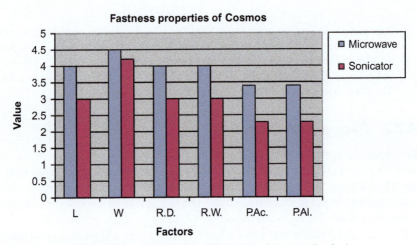

Fig. 4A.13 Dye uptake in microwave dyeing with cosmos (post-mordanting).

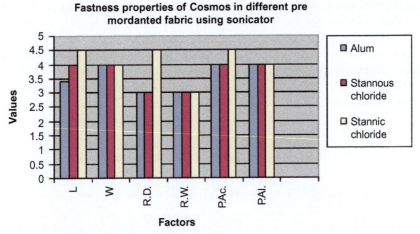

Fig. 4A.14 Comparative fastness properties of cosmos in different pre-mordanted cotton.

while least in the case of alum. Washing fastness is same for $SnCl_2$, $SnCl_4$, and alum. Wet rubbing fastness is also the same for all the three mordants but dry rubbing fastness shows best result for $SnCl_4$, however $SnCl_2$ and alum show lower results. Similarly alkaline perspiration fastness is the same with all the three mordants but in case of acidic perspiration $SnCl_4$ is better than alum. The findings of color fastness tests of cotton samples to rubbing under dry conditions showed that the samples had good to excellent fastness to dry rubbing. The samples when subjected to wet rubbing exhibited a decrease in color fastness rating than dry rubbing representing fair to good fastness. Similarly color fastness of samples to perspiration, showed that acidic perspiration had better fastness results as compared to alkaline perspiration tests. Washing fastness with Cosmos dyeing on cotton fabric postmordanted with stannous chloride shows the best result of all the mordanting methods (Vankar et al., 2001). Table 4A.13 shows the shade

Table 4A.13 **L*, a*, and b* values for cotton**

Method premordanting	Mordant	L*	a*	b*	Color obtained
	Cosmos dye	−40.65	12.50	24.38	Bright orange
	FeSO$_4$	−58.50	14.2	30.55	Dark olive green
	Alum	−56.28	13.52	28.55	Brownish yellow
	SnCl$_4$	−36.45	14.55	32.25	Light mustard
	SnCl$_2$	−42.55	12.67	25.85	Brownish orange

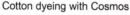

Cotton dyeing with Cosmos

Fig. 4A.15 Shade card of cotton dyed by *Cosmos*.

constants of the cosmos dyes as well as the fabric dyed using different mordants and Fig. 4A.15 gives an idea about various shades of yellow and orange by *Cosmos*.

Cosmos shows good fastness properties on cotton with different mordants. A variety of shades is obtained. This flavone based chromophore is relatively stable to photo fading and the dye may be present in an aggregated form inside the cotton fabric due to which it is fast to daylight. Washing fastness also shows excellent results. Thus cosmos dye exhibits good affinity for cotton by the use of ecofriendly mordants and is a potential natural colorant for the dyers.

4A.7   Dyeing cotton with *Terminalia arjuna*

4A.7.1   *Standardization, optimization of* Terminalia arjuna *dye*

Optimization of extraction conditions include some important parameters like extraction time, extraction temperature, pH of extraction medium, and mass to liquor ratio. The extracted dye has been studied for dyeing using different mordants for cotton fabric and their L, a*, and b* values. The washing, rubbing, light, and perspiration fastness of the dyed samples were also evaluated, giving fair to good fastness grades. The dye and the dyed fabric have been studied for eco-friendliness.

4A.7.2   *Extraction and purification of dye*

pH of extraction medium: The raw material (*Terminalia arjuna* bark) was grounded. The ground bark (10 g each) was soaked in beaker containing water (app. 200–250 mL) of different pH (4, 7, and 9). It was filtered after cooling through ordinary filter paper and the filtrate was collected, dried at $70 \pm 5°C$ and weighed to calculate % yield of the extracted mass.

Mass to liquor ratio The ground bark was soaked in four beakers having 100, 200, 300, and 400 mL at 70°C for 3 h. It was filtered through ordinary filter paper and the filtrate was collected, dried at $70 \pm 5°C$ and weighed to calculate % yield of the extracted mass.

Solubility: Water, acetone, ethanol, hexane, methanol, and ethyl acetate were used for ascertaining the solubility of extracted dye. About 0.5 g of the extracted dyes was taken in different conical flasks and 20 mL of each solvent was added in separate conical flasks.

4A.7.3   *Dyeing*

For dyeing, sonicator has been used as dye bath. Extracted dye is taken in sonicator bath and treated fabric is dipped in it for 1 h. After one hour it is dried in shade. Dye uptake by the fabric is monitored by the lowering of optical density of the dye bath solution and also by the shade of the color that appears on the fabric. The tone can be adjusted as per the requirement. Same method has been followed for conventional dyeing. Dyed fabric is dipped in 4% sodium chloride solution for one hour and then fabric is washed with tap water and dried in the shade. Aqueous extract of *Terminalia arjuna* give different shades of brown with different mordants as shown in Table 4A.14 (Tiwari and Vankar, 2007).

Table 4A.14   **Color obtained and L*, a*, and b* values for dyed cotton fabric with terminalia**

Mordant (premordanting)	Color obtained	L*	a*	b*
Alum	Coffee brown	43.021	37.307	27.033
Stannic chloride	Skin color	81.000	11.030	23.100
Stannous chloride	Light brown	62.010	17.100	30.003
Ferrous sulfate	Blackish coke	31.100	23.000	−2.000
Copper sulfate	Brown	55.100	20.030	27.005
Pot. dichromate	Dark brown	47.000	21.500	27.100

Table 4A.15 fastness properties for cotton fabric dyed with terminalia

| Mordant (premordanting) | Washing IS-687-79 | Light IS-2454-85 | Rubbing IS-766-88 | | Perspiration IS-971-83 | |
			Dry	Wet	Alkaline	Acidic
Alum	4	III	4–5	4	4/4	4/4
Stannic chloride	4–5	V	4–4	4	4	4
Stannous chloride	4–5	IV	4/5	4	4	4
Ferrous sulfate	4	V	4/4	4	4	4
Copper sulfate	4–5	V	4	4	4	4
Pot. dichromate	4–5	IV	4	4	4	4

Washing, light, and rubbing fastness is very good for *Terminalia arjuna*. Microchemical analysis showed that extracted dye gives violet red color with hot ethanol confirm the presence of anthraquinone. As the main colorant in *Terminalia* is esters of triphenic acids in all the fastness parameters it shows good result, as tannins adhere to cotton well, pretreatment with metal mordants such as alum, potassium dichromate, copper sulfate, ferrous sulfate, stannic chloride, and stannous chloride shows fastness even better than conventional tannic acid pretreatment. The fact that the washing and light fastnesses of these six different mordants are compared, shows that the fastness properties in most of the cases is greater than 4, it makes *Terminalia arjuna* a potential entrant in textile industry. *Terminalia* can also be extended for dyeing wool in carpet industry as reddish brown is a very good color (Table 4A.15).

4A.8   Ultrasound energized dyeing of cotton by Tulsi leaves (*Ocimum sanctum*) extract

4A.8.1   Dye material

A fragrant bushy perennial generally found in every house. This herb is held sacred by the Hindus and is often planted around temples, also used in rosaries. However nobody has ever tried to use it for dyeing. This herb is used for dyeing for the first time. Dye was extracted in methanolic solution using a soxhlet. The extraction was carried out for 3–4 h. The solution was then evaporated to half the original volume. The methanolic extract of leaves is a dark green solution with a characteristic odor. The flavones apigenin and luteolin, the flavone-7-O-glycoside, luteolin-7-O-glucuronide, and the flavone C-glucosides are the main colorants (Grover and Rao, 1997).

4A.8.2   Fabric and mordants

Plain weave, white cotton broad cloth, which is generally used for dyeing was selected. The mordants used for study were Alum, Stannous chloride, Stannic chloride, and ferrous sulfate.

The fabric was desized in liquor containing 5 g of soap and 0.1% HCl/L of water. The material to liquor ratio was taken as 1:40. The fabric was boiled at 95°C for one hour and rinsed thrice in cold water, and dried under shade. The desized cotton fabric was treated with tannic acid solution. The material to liquor ratio was 4% (owf). The fabric was soaked in the tannic acid solution for 4–5 h and then air dried.

4A.8.3 Dyeing with Tulsi leaves

Dyeing with Tulsi leaves does not give much variation in color. However color adherence to fabric is good. With Stannic chloride, stannous chloride, and alum light green, fluorescent green and green shades are obtained respectively. Postmordanting of stannic chloride is better where as stannous chloride and alum gives better results with premordanting as shown in Tables 4A.16 and 4A.17. With ferrous sulfate dark green Khaki color is obtained, premordanting is better but color is not uniform in either of the cases. Tulsi leaves dye is good in terms of color fastness. In general we can conclude that Tulsi leaves dye exhibit fair to good fastness to rubbing and good fastness to light and washing. Hence this dye is well suited for cotton which is subjected to laundering more often than their synthetic counterparts. Cost effectivity of dyeing was also verified and found that the cost in dyeing process depended mainly on the cost of mordants (Tiwari et al., 2000). As the use of sonicator is for more economical, dye-uptake eventually works out to be cheaper.

A current trend in the processing of textiles is to combine several different finishes in one bath, resulting in several different final fabric characteristics. However in this

Table 4A.16   Fastness grades of Tulsi dyed cotton

	Light fastness	Wash fastness	Rubbing fastness
With mordant	4/5	4/5	5
Without mordant	3/4	4/5	2/3

Table 4A.17   Fastness grades of Tulsi dyed cotton (pre and post mordanting)

	Dyed shade			Fastness properties				
						Rubbing		Perspiration	
Mordant	Pre	Post	Wash	Light	Dry	Wet	Alkaline	Acidic
Stannic chloride	Light green	Light green	4/5	4–5	4	3/4	4/5	4/5
Stannous chloride	Fluorescent green	Light green	3/4	4	3	3/4	4	4
Ferrous sulfate	Khaki green	Khaki dirty green	4	4	3–4	3–4	4	4
Alum	Green	Geen	4	4	3/4	3–4	3–4	3–4

case, it is one dip, additive added methodology, whereby dyeing and antimicrobial finish together. The leaves of Tulsi yield an essential oil, which contains eugenol, carvacrol, methyl eugenol, limatol, and caryophyllene. The leaves also contain Ursolic acid, aspigenium, luteolin, apigenium-7, glucuronid, orientin, and molludistin. Ursolic acid ($C_{30}H_{48}O_3$, mol. wt 456.71) is the active ingredient present about 75% w/w. ß-Ursolic acid (triterpenoid sapogenin from the ursan group inhibited the growth of several strains of staphylococci. Numerous ursolic acid containing plants from the Lamiaceae family exhibit antibacterial/fungal activity (Liu, 1995).

The other components are all volatile oil content, so the leaves are fast dried below 110°C so that the volatile content is not lost. Color, flavor, and antimicrobial properties are preserved by fast drying. The Tulsi leaves have long shelf life. On storage the leaves show no deterioration of either the rich green color nor in the antimicrobial activity. The effectiveness of the antimicrobial finish is not inhibited by dyeing nor is it influenced by it. Since the antimicrobial agent and the green pigment are derived from natural source they behave in a complimentary manner. No chemical interaction in the solution occurred in the sonicator and processing time and steps are minimized as compared to multiple treatments.

Scanning Electron Microscopy (SEM): Evaluation by Scanning Electron Microscopy (SEM) of the untreated and treated material high resolution images of the morphology or topography of the fabric is obtained. Compositional analysis of the material is also obtained by monitoring secondary X rays produced by the electron- specimen interaction. The study is made to observe the surface change, such as a presence and uniformity of additive, before and after application of the Tulsi extract. The untreated sample of the fabric had an uneven surface when viewed using SEM. However, the surface of the treated fabric was very smooth and distinct due to the high add-on level of deposits from Tulsi extracts as can be seen in the SEM picture (Fig. 4A.16).

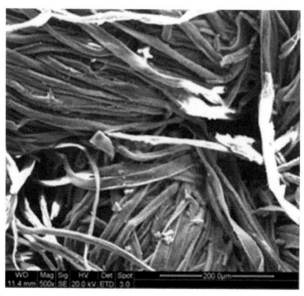

Fig. 4A.16 SEM of the Tulsi dyed cotton fabric.

Evaluation of the dyeing properties reveal that mordanting the fabric in the tannic acid prior to dyeing and then postmordanting with the stannic chloride show reasonably good fastness properties. The fabric showed moderate dyeing effectivity without the usage of mordants.

Estimation of antimicrobial property: The untreated and treated fabrics were both subjected to soil burial test for three days. The breaking strength of the fabric was then evaluated. Treated fabric showed very encouraging results.

The dyeing and the antimicrobial treatment achieved by methanolic extract of Tulsi leaves, in one bath by using sonicator shows very effective dyeing as can be evaluated by the fabric's fastness properties. Similarly the antimicrobial activity evaluated by soil burial tests of both treated and untreated fabrics show the resistance to microbial growth on the treated fabric after soil burial for 3 days. The environmental activists are supportive of using natural colorants as they are seen to be using renewable resources, causing minimum pollution, and having less risk to human health. There is currently a demand for natural dyes in a niche market that could be expanded. The large scale production of textiles dyed with natural dyes is a new concept for the textile industries. Tulsi leaves dye extract will definitely find great use in cotton industry especially in green color range dyeing.

4A.9   Ecofriendly sonicator dyeing of cotton with *Rubia cordifolia* Linn. using biomordant

4A.9.1   Dye material

Dye used was *Rubia cordifolia* and biomordant *Eurya acuminata* DC. var euprista Karth (Nausankhee, Turku, belonging to the Theaceae family) was used. Dyed cotton has shown very good fastness properties using dry powder as 10% of the weight of the fabric. Use of biomordant replaces metal mordants making natural dyeing ecofriendly.

Only a small fraction of plant species take up high levels of aluminum (Al) in their above-ground tissues. Generally, plants are classified as accumulators if they accumulate at least $1000\,mg\,kg^{-1}$ in their leaves (Chenery, 1948) (Robinson and Edgington, 1945). Knowledge of Al accumulators is built mainly on the substantial contributions made by researchers in the last 50 years or so. The extract of *Eurya acuminata* DC var euprista Karth leaves is found to contain substantial amount of Al. Aluminum accumulation is a primitive character mainly characteristic of woody and tropical representatives of fairly advanced families (e.g., Anisophylleaceae, Hydrangeaceae, Melastomataceae, Rubiaceae, Theaceae, Symplocaceae, Vochysiaceae) (Chenery and Sporne, 1976) (Figs. 4A.17 and 4A.18).

Dyeing with rubia was carried using biomordants and a comparative study was also undertaken with metal mordant (Vankar et al., 2008b). The CIE Lab values were ascertained with and without biomordant for *Rubia cordifolia* as given in Table 4A.18.

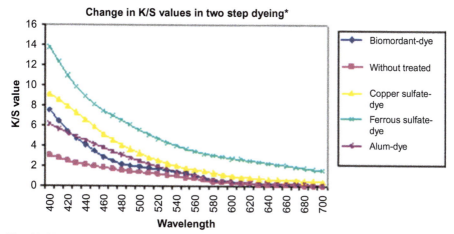

Fig. 4A.17 Pre-mordanting with metal and bio-mordant for cotton dyed by *Rubia cordifolia*.

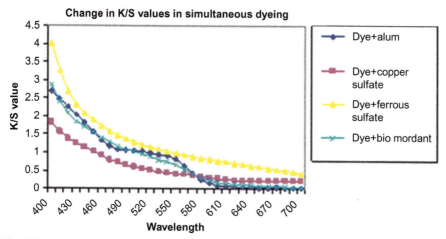

Fig. 4A.18 Simultaneous mordanting with metal and bio-mordant for cotton dyed by *Rubia cordifolia*.

Table 4A.18 **Effect of *E. acuminata* (Nausankhee) biomordant and metal mordant (alum) on the colorimetric data obtained for *Rubia cordifolia* dye**

Experiment	L*	a*	b*	C	H
Without mordant	59.851	15.246	23.401	27.929	56.892
With biomordant	59.915	19.560	32.493	37.926	58.929
With alum mordant	64.372	17.043	34.088	38.111	63.411
With CuSO₄ mordant	63.685	7.069	31.338	32.125	77.257
With Fe SO₄ mordant	59.613	4.700	21.879	22.378	77.845
With biomordant + dye	68.130	18.129	27.387	32.844	56.474
With alum mordant + dye	61.848	20.774	26.726	33.850	52.121

4A.9.1.1   Effect of mordanting conditions

It was observed that the premordanting technique with biomordant and metal mordants imparted better fastness properties to the cotton fabric compared to simultaneous mordanting techniques. Therefore, premordanting technique was applied; the dyed fibers were mordanted with biomordant, $CuSO_4$, $FeSO_4$, and Alum.

The mordant activity of the four cases followed the sequences

$Fe \rightarrow Cu \rightarrow Al \rightarrow$ biomordant in cotton for *Rubia cordifolia*, the absorption of color by cotton fabric was good for both metal mordants and biomordant. This might be due to the maximum absorption and easy formation of metal-complexes with the fabric.

4A.9.1.2   Dyeing and fastness tests

Dried stems, leaves, and roots of *Rubia* ground into a fine powder and used to dye cotton, using a sonicator dyeing apparatus (Julabo, Germany). The fabric to be colored was previously washed with mild soap solution (Labolene) at 40°C for 30 min. They were then mordanted with 2% (weight/commercial weight) of *Eurya acuminata* DC. var euprista Karth at 40°C for 60 min. The ground powder (30% weight/commercial weight) was mixed with water (powder–water = 1:5), stirred for 60 min at the temperature of 40°C and then filtered. This extract was used for filling the dyeing system, in which the substrate was placed. The dyeing process took place at 30°C for 60 min and at atmospheric pressure. Finishing consisted of several rinsing with water.

4A.9.1.3   Fastness testing

The dyed samples were tested according to Indian standard methods. The specific tests were: color fastness for light, IS-2454-85 by a xenotest to measure resistance to fading by using a laboratory apparatus (Xenotester- Alpha, Germany) under the following conditions: light-exposure system featuring an air-cooled Xenon arc discharge lamp simulating outdoor global radiation; irradiation on sample level λ 300–400 and 400–700 nm; test chamber temperature: 25°C; relative humidity 65%. A light fastness rating from 1 (severely fading) to 8 (fast) was made by comparing the resistance to fading of each sample to that of eight different blue tones, color fastness to rubbing, *IS-766-88*, color fastness to washing *IS-687-79*, wash fastness rating from 1 (fading) to 5 (excellent fastness) was used, color fastness to perspiration, *IS-971-83* (Table 4A.19).

Table 4A.19   **Fastness properties of dyed cotton fabrics under conventional heating and ultrasonic conditions of biomodanting and *R. cordifolia***

Dyeing methods	Wash–perspiration–rubbing–light					
	WF	Per$_{acidic}$	Per$_{basic}$	Rub$_{dry}$	Rub$_{wet}$	LF
Conventional	4	4	3–4	3–4	3–4	4
Ultrasonic	5	4–5	4–5	4	4	4–5

WF = wash fastness, LF = light fastness. Light fastness accordingly : 1, severely fading; 2, fading; 3, fairly fading; 4, quite good fastness; 5, good fastness; 6, very good fastness; 7, excellent fastness; 8, exceptional fastness; Wash fastness accordingly: 1, fading; 2, fairly fading; 3, medium fastness; 4, good fastness; 5, excellent fastness.

Fig. 4A.19 Shade card of *Rubia cordifolia* (different concentration).

Dyeing with *Rubia* was carried out with different concentrations (Fig. 4A.19). The one stage and the two stages dyeing of cotton fabric with and without bio-mordant by the natural dye *Rubia cordifolia*, show that two stage processes with biomordant provided very good results. The dye uptake in case of two step dyeing is 14.8%, 23.5%, and 33.5% for without mordant, biomordant, and alum mordant. In the case of one step dyeing the dye uptake is 38% and 47% for dye-biomordant and dye-alum simultaneous mordanting methods. The effectiveness of biomordant-*R. cordifolia* in better dye uptake may appear to be slightly less as compared to metal mordanting however the reduction in effluent pollution as well as improved fastness properties outweigh its benefit as observed during the course of this study. The pH of *Rubia cordifolia* extract is 5.7 where as the pH of *Eurya acuminata* DC. var euprista Karth. (nausankhee) extract is 7.67. Thus it can be said that the two extracts are complimentary to each other and that causes the better dye adherence which gave beautiful shades of pink (Figs. 4A.20 and 4A.21). The suitability of specific biomordant *E. acuminata* (nausankhee) for this particular natural dye was evaluated on the basis of the traditional information collected from the tribal people.

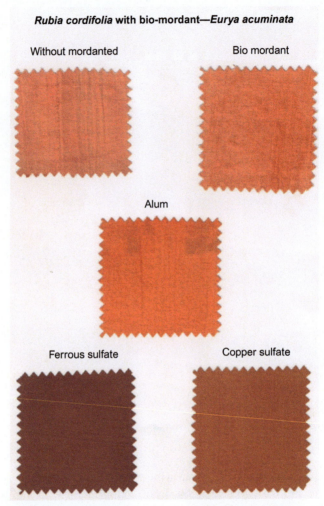

Fig. 4A.20 Shade card of *Rubia cordifolia* with biomordant (*E. accuminata*).

Rubia cordifolia was found to have good agronomic potential as a dye crop in Arunachal Pradesh. Biomordant *Eurya acuminata* DC. var euprista Karth (Nausankhee) when used in conjunction with *Rubia cordifolia* was found to enhance the dyeability due to the Al contents present in the leaves. Enhancement of dye uptake was 23.5% with biomordant, 33.5% with alum, and 14.8% without any mordant. Use of bio-mordant not only enhances the fastness properties but it also gives good colorimetric data on dyeing. Even the fastness properties in this case show good results. The two step biomordant-dye is developed for the ease of industrial application and offers an ecofriendly process which should be popularized as an alternate method to the metal mordant-dye method.

Rubia cordifolia with bio-mordant—*Eurya acuminata*

Fig. 4A.21 Shade card of *Rubia cordifolia* (*E. accuminata*).

Conclusion

In the present context of eco-preservation, natural dyes have gained incredible commercial prospectives. The revival of significance in the use of natural dyes in textile coloration has been gaining incessant popularity all over the world, probably due to environmental concerns, eco-safety, and pollution control. Commercialization of natural dyes can be done successfully by a systematic and scientific approach to extraction, purification, and use of natural dyes. Optimization of extraction conditions is a must to minimize the investment cost and to avoid discrepancy in the dyed shade quality. Natural dyes are safe and eco friendly as it has been shown to be free from hazardous chemicals and play an important role in cotton dyeing.. Therefore, its commercial use shall definitely minimize the health hazards caused by the use of synthetic dyes.

Further Reading

- M.J., Melo, 2009. History of natural dyes in the ancient mediterranean world. *Handbook of natural colorants*, pp. 3–20.
- P.S., Vankar 2013. Handbook on Natural Dyes for Industrial Applications *(with Color Photographs)*. National Institute of Industrial Research. New Delhi-110007, India.
- D. Cristea and G., Vilarem, 2006. Improving light fastness of natural dyes on cotton yarn. *Dyes and pigments*, *70*(3), pp. 238–245.

- A.K. Samanta, and P. Agarwal, 2009. Application of Natural Dyes on Textiles.
- D. Cardon, 2007. Natural dyes: sources, tradition, technology and science. Archetype.
- İşmal, Ö.E., Yıldırım, L. and Özdoğan, E., 2014. Use of almond shell extracts plus biomordants as effective textile dye. Journal of Cleaner Production, 70, pp. 61–67.

References

Chenery, E., 1948. Aluminium in plants and its relation to plant pigments. Ann. Bot. 12, 121–136.

Chenery, E.M., Sporne, K.R., 1976. A note on the evolutionary status of aluminium accumulators among dicotyledons. New Phytol. 76, 551–554.

Ghorpade, B., Tiwari, V., Vankar, P.S., 2000. Ultrasound energised dyeing of cotton fabric with canna Flower extracts using eco friendly mordants. Asian Text. J. 68–69.

Grover, G.S., Rao, J.T., 1997. Investigations on the antimicrobial efficiency of essential oils from *Ocimum sanctum* and *Ocimum gratissimum*. Perfum. Kosmef. 58, 326–328.

Liu, J., 1995. Pharmacology of oleanolic acid and ursolic acid. J. Ethnopharmacol. 49, 57–68.

Robinson, W., Edgington, G., 1945. Minor elements in plants, and some accumulator plants. Soil Sci. 60, 15–28.

Srivastava, J., Seth, R., Shanker, R., Vankar, P.S., 2008. Solubilisation of red pigments from Canna Indica flower in different media and cotton fabric dyeing. Int. Dyer, 31–36.

Tiwari, V., Vankar, P.S., 2007. Standardization, optimization and dyeing cotton with *Terminalia arjuna*. Int. Dyer 31–33, 35–36.

Tiwari, V., Ghorpade, B., Mishra, A., Vankar, P.S., 2000. Ultrasound dyeing with *Ocimum sanctum* (Tulsi leaves) with ecofriendly mordants. New Cloth Market 23.

Tiwari, V., Shanka, R., Vankar, P.S., 2003. Ultrasonic dyeing cotton fabric with Babool bark (*Acacia arabica*) for preparation of eco-friendly textile. Chem. World 30–32.

Vankar, P.S., Kushwaha, A., 2011. Salvia splendens, a source of Natural dye for cotton and silk fabric dyeing. Asian Dyers 29–32.

Vankar, P.S., Shanker, R., 2009. Potential of *Delonix regia* as new crop for natural dyes for silk dyeing. Color. Technol. 125, 155–160.

Vankar, P.S., Shanker, R., 2010. Natural dyeing of silk and cotton by Rhododendron flower extract. Int. Dyers, 37–40.

Vankar, P.S., Srivastava, J., 2010. Ultrasound-assisted extraction in different solvents for phytochemical study of *Canna indica*. Int. J. Food Eng. 6.

Vankar, P.S., Tiwari, V., Ghorpade, B., 2001. Microwave dyeing of cotton fabric-*Cosmos sulphureus* and comparison with sonicator dyeing. Can. Text. J. 31.

Vankar, P.S., Shanker, R., Dixit, S., Mahanta, D., Tiwari, S., 2008a. Sonicator dyeing of modified cotton, wool and silk with *Mahonia napaulensis* DC. and identification of the colorant in Mahonia. Ind. Crop. Prod. 27, 371–379.

Vankar, P.S., Shanker, R., Mahanta, D., Tiwari, S., 2008b. Ecofriendly sonicator dyeing of cotton with *Rubia cordifolia* Linn. using biomordant. Dyes Pigments 76, 207–212.

Dyeing application of newer natural dyes on silk with fastness properties, CIE lab values, and shade card

D. Shukla, P.S. Vankar
FEAT (Facility for Ecological and Analytical Testing), Kanpur Kalyanpur, India

Introduction

Silk dyeing with the plant extracts has been practiced since ancient times, however exploring newer natural dye sources has been a recent phenomena. Silk being a protein material binds to colorant molecules more strongly than cotton and its own shine adds luster to the colored fabric. Like cotton, silk also has to be prepared for dyeing by usual method of scouring.

4B.1 Silk dyeing with extract of Black carrot

4B.1.1 Preparation of the fabric

Silk fabric was washed with a solution containing 5 g/L of sodium carbonate and 3 g/L of nonionic detergent (Labolene) for 1 h, after which it was thoroughly rinsed with water and air dried at room temperature.

4B.1.2 Dyeing silk with Black carrot

After mordanting, dyeing with aqueous extract of Black carrots or *Daucus carrota* was carried out using approximately 6% of dye solution. Dyeing was performed using a liquor ratio of 1:20 at 65°C in open beakers with manual agitation of the extract. Mordanted silk pieces were dipped in dye bath for 3 h, and then they were taken out, dried, and fixed (Shukla and Vankar, 2013).

4B.1.3 Shades obtained on silk by Black carrot

The pigments (anthocyanin) present in black carrots color exhibit a reversible change in molecular structure as the pH of solutions change from acidic to basic. This change in structure is characterized by a shift in hue from red to black to blue as the pH changes from acidic to basic. In the Table 4B.1, obtained color varies from light green

Natural Dyes for Textiles. http://dx.doi.org/10.1016/B978-0-08-101274-1.00010-0

to olive. Dyeing on silk showed acceptable fastness to light and water for all mordant variations, although the dyeing with use of Stannous and Iron mordant exhibited sufficient high fastness to be of interest for textile dyeing. Shade and color depth on silk with use of metal mordants was of significant importance for a color gamut based on this natural dye, i.e., black carrot. Two enzymes viz. protease and trypsin as mordants did not show any improvement in color strength and shades and thus was not considered for further analysis. The anthocyanin of black carrot do not show much affinity to enzyme mordanted silk as it showed in the case of wild blue berry/*Cayratia carnosa* (Vankar and Shanker, 2006a,b; Shanker, 2011).

4B.1.4   CIE La*b* values of the dyed silk

CIE La*b* values for silk dyed with different metal mordants are given in Table 4B.1. All the mordants have shown almost equal lightness in the range 54–56. Stannous mordanted silk was best among them with lightness scale of 54.23 and Iron mordanted silk with 54.91. Application of these metallic salts in ecologically safe manner/amount can give very good shades with black carrot. The fastness properties are shown in Table 4B.2.

Table 4B.1   **CIE La*b* values and shades obtained for silk dyed with extract of Black carrot**

S. no.	Mordant	Shades obtained	L	a*	b*	K/S
1.	Control		55.85	−1.69	3.74	23.31
2.	Alum		55.20	−6.67	−3.58	40.93
3.	Copper sulfate		55.48	−8.22	−1.89	37.46
4.	Ferrous sulfate		54.91	−5.56	−2.54	130.21
5.	Pot. dichromate		55.86	−7.33	0.55	27.00

Table 4B.1 **Continued**

S. no.	Mordant	Shades obtained	L	a*	b*	K/S
6.	Stannous chloride		54.23	2.63	−5.96	78.31
7.	Stannic chloride		55.56	1.53	2.21	44.31

Table 4B.2 **Fastness properties for silk dyed with Black carrot**

	Fastness properties					
			Rubbing IS-766-88		Perspiration IS-971-83	
Premordanting	Washing IS-687-79	Light IS-2454-85	Dry	Wet	Alkaline	Acidic
Alum	4–4/5	IV	4–5	4–5	4/5	4/5
Copper sulfate	4–4/5	III	3–4	3	3–4	3–4
Ferrous sulfate	4–4/5	IV	3–4	3–4	3–4	3–4
Stannous chloride	4–4/5	III	4	3–4	3/4	4
Stannic chloride	4–5	IV	4	4	4	4
Pot. dichromate	4–4/5	IV	4	4	3/4	4

Black carrot as a dye source was tested for its color compatibility with silk fabric with premordanting process which helped to enhance the color depth and dye ability. It has substantial amount of anthocyanin as primary colorant acting as good dye source for silk giving very beautiful earthy shades of green. As black carrot has given good shades on silk (being a protein fiber), can be another good preposition. Acceptance of this new dye source of black carrot can enhance an opportunity for new markets and new businesses.

4B.2 Silk dyeing with anthocyanins from *Hibiscus rosa sinensis* flowers

Anthocyanins from *Hibiscus* flowers have been extracted using novel technique by using citric acid with methanol instead of hydrochloric acid for textile dyeing particularly silk using different mordants (Vankar and Shukla, 2011). A new approach for natural dyeing with anthocyanin has been discussed along with a convenient method of extraction of anthocyanin from Hibiscus flowers, which has been developed using methanolic solution of 4% citric acid. The new method gave better yield of anthocyanin as compared to methanolic solution of 0.1% hydrochloric acid. It has been also shown that pH of the extract plays an important role on the dye, thus by adjusting the pH of the extract at 4, dyeing of silk together with metal mordanting gave different colors. The best dyeing results were obtained for stannous mordanted fabrics in terms of fastness properties.

4B.2.1 Dye material

Hibiscus rosa sinensis flowers were collected from the Indian Institute of Technology Campus, Kanpur, India and they were kept in cold (20°C) and dark storage until processed (the shelf life was found to be for more than a month). Petals of flowers chosen were cut into small pieces and extracted into methanol (96% (v/v)), keeping them overnight.

Anthocyanins are one of the most abundant natural pigments available. These are the vacuolar pigments found in almost every part of higher plants and water soluble strong colors and have been used to color food since historical times (Welch et al., 2008). Chemically, anthocyanins are subdivided into the sugar-free anthocyanidine aglycons and the anthocyanin glycosides. As vegetative dyestuff must have oxochrome groups in order to obtain good results in dyeing, anthocyanin has many oxochromes. Cyanidin has five oxochrome groups. The Hibiscus anthocyanin mainly comprises of Cyanidin-3-Sophoroside (Delgado-Vargas et al., 2000).

The stability of anthocyanins depends on the structural changes between flavylium cation. The color of the extract obtained was also affected by the pH value in the solution ranging from red to dark purple at pH 2.5 to pH 5. However, the total anthocyanin content in the extracts was quite stable in the pH values ranging from 2.5 to 5.0 used in the study. The anthocyanin content was stable at lower pH (<3) but the color of the extracts faded at higher pH values (<4.5). Degradation percentages of total anthocyanins in the extracts kept at 25°C were 7%–20% lower than that maintained at 35–40°C. The study shows that suitable storage condition for colored anthocyanin pigments in extracted form is under acidic conditions and should be kept in the dark. Other factors found to affect the pigment stability were light and elevated temperature which caused increased pigment degradation. Fig. 4B.1 shows color of anthocyanin at different pH.

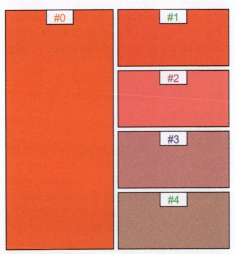

Fig. 4B.1 Color of anthocyanin extract at intrinsic and different pH (#0–4 in order of intrinsic pH, pH 1, pH 2, pH 3, and pH 4).

Table 4B.3 Color coordinates of the hibiscus extract at different pH

	Intrinsic 2.55	pH 3	pH 4	pH 5	pH 6
L	45.78	20.39	6.75	2.79	1.93
a*	66.14	50.16	34.86	19.27	13.64
b*	58.50	35.02	11.62	4.79	3.33
C	88.30	61.17	36.74	19.85	14.04
H	41.47	34.90	18.42	13.97	13.72

4B.2.2 Dyeing of silk by Hibiscus anthocyanin extract

Table 4B.3 shows the CIE La* b* values of hibiscus extract at different pH values, while Table 4B.4 shows dyed silk fabric with Hibiscus anthocyanin extract after pretreatment with different metal mordants, the dyeing with different mordants imparted a shade change from pink, brown to purple. Varied hues of color were obtained by premordanting the silk with alum, $SnCl_2$, $CuSO_4$, and $K_2Cr_2O_7$ and were dyed by anthocyanin extract of hibiscus flowers as shown in the Fig. 4B.2 and Table 4B.4. The different mordants not only cause difference in hue color and significant changes in K/S values but also L values and brightness index values. The best values are obtained with stannous chloride.

Table 4B.4 Silk dyeing with hibiscus anthocyanin

Mordant	L	a*	b*	K/S
Control	43.45	31.40	−0.17	52.34
Alum	43.48	31.03	0.04	51.14
$CuSO_4$	43.34	30.10	−0.02	48.31
$K_2Cr_2O_7$	43.45	27.41	2.41	66.53
$SnCl_2$	40.54	16.27	−14.08	128.88

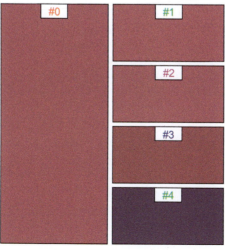

Fig. 4B.2 Hibiscus dyed silk with different mordants (#0–4 in the order of control, alum, copper sulfate, potassium dichromate, and stannous chloride).

In the dyeing of silk fabrics, the best results were obtained from the premordanting method at acidic value (pH-4). Different color shades/tones were obtained of pinks (Al, Cu and Cr) and purple (Sn). These color tones were obtained at pH 4. Utilization of anthocyanin extract from hibiscus flowers has been found to have good agronomic potential as a dye plant. Metal mordant when used in conjunction with the anthocyanin extract of *Hibiscus rosa sinensis* was found to enhance not only the dye ability but also the fastness properties of the dyed fabrics as compared to the controlled sample. Enhancement of dye uptake was better than nonmordanted fabric. Even the fastness properties in this case showed good results. The two step process of premordanting and then dyeing was developed for the ease of industrial application.

Table 4B.5 Fastness properties of dyed cotton and silk fabrics with different metal modants with hibiscus

Dyeing methods	Wash-perspiration-rubbing-light					
	WF^a	Per$_{acidic}$	Per$_{basic}$	Rub$_{dry}$	Rub$_{wet}$	LF^b
Silk (control)	2	2	2–3	2–3	2	2
Silk (Alum)	4	4	4	4	4	4
Silk (SnCl$_2$)	5	4–5	4–5	4–5	4–5	4–5
Silk (CuSO$_4$)	4–5	4	4	4	4	4–5

^{a} Wash fastness.
^{b} Light fastness.

The results of the washing fastness were also investigated and it was seen that the results were good particularly for SnCl$_2$ mordanted fabrics as shown in Table 4B.5 due to special conjugation of anthocyanin color moieties and stannous.

Hibiscus extract, already used to give color and flavor to beverages and many other food items whereas utilization of anthocyanin extract from hibiscus flowers as a source of dye has been attempted for the first time for textile dyeing. Use of metal mordants such as Sn, Al, and Cu in conjunction with the anthocyanin extract of *Hibiscus rosa sinensis* was found to enhance the dye ability along with improved fastness properties of the dyed fabrics as compared to the controlled sample. The two step dyeing process of cotton and silk fabrics with premordanting method at acidic value (pH 4) yielded different color tones on silk and cotton specially one with tin mordant having very good wash and light fastness. The developed shades will surely be liked by consumers in the present global textile market.

4B.3 Silk dyeing with *Delonix regia*

This is the first description where *Delonix* flowers extract has been used as a source of natural dye for silk dyeing in conjunction with enzymes- Protease, amylase, diasterase, and lipase and the source of biomordant is *Pyrus pashia*. The three main objectives of the present study using the extract of *Delonix* for the dyeing of silk fabrics are:

- To find an alternative to metal mordanting, making the dyeing process ecofriendly
- To check the compatibility of enzymes and biomordant with *Delonix* dye as both enzyme and biomordant are known to be substrate specific
- To carry out the dyeing process under mild operating conditions

The objective of this study was to find out the utilization of enzymes and a biomordant for silk dyeing with *Delonix* natural dye and to replace metal mordants. For the evaluation of better dyeing results two different strategies were adopted such as (a) One step process with meta-mordanting/simultaneous mordanting and (b) Two step process with premordanting followed by dyeing. This was carried out separately for each enzyme such as Diasterase, Lipase, and Protease-Amylase. Similarly experiments were carried out for biomordant- pyrus as well as for metal mordants- alum, copper sulfate, and ferrous sulfate, and lastly to compare the dyed swatches with metal mordanted swatches.

4B.3.1   Dyeing with Delonix regia

The objective of this study was to find out the utilization of enzymes and a biomordant for silk dyeing with *Delonix* natural dye and to replace metal mordants. The effect of different enzymes and a biomordant for checking the dye /enzyme or dye/biomordant suitability has been observed. For the evaluation of better dyeing results two different strategies were adopted such as (a) One step process with meta-mordanting/ simultaneous mordanting and (b) Two step process with premordanting followed by dyeing. This was carried out separately for each enzyme such as Diasterase, Lipase, and Protease-Amylase. Similarly experiments were carried out for biomordant-pyrus as well as for metal mordants-alum, copper sulfate and ferrous sulfate, and lastly to compare the dyed swatches with metal mordanted swatches (Vankar et al., 2007b; Vankar and Shanker, 2008).

The fabric pretreated with biomordant showed good dye adherence, this prompted us to carry out analysis of the metal content in the biomordant source *Pyrus pashia* fruit. Atomic Absorption Spectroscopy and the analytical results show the presence of copper (Cu) in 10.66 mg/100 g. It is present in some chelated form therefore it might help the dye to adhere to the fabric similar to metal mordanting effect. The high Cu content suggested stronger and useful chelation to the colorant for better dye adherence. The presence of the 4-oxo group in quercetin in conjunction with hydroxyl group also helps in chelation in flavonoids. Chelation of copper on the site between the 4-oxo group and C-5 OH group in flavonols and flavones has already been proposed (Mira et al., 2002; Brown et al., 1998). The number of OH groups is also important, the higher the number, the higher their chelating ability is. Thus in the similar manner, the copper in Pyrus helps in chelation of Cu(II) with the colorant comprising mostly of flavones/flavonols. Based on the above literature precedence the probable mode of chelation of copper in *Pyrus* with quercetin has been proposed. Chemical binding of the dye with biomordant has also been proposed analogous to metal-dye complex formation.

4B.3.2 Evaluation of enzyme/biomordant treated dyed fabrics for change in K/S and color strength

The results of K/S and color coordinate values have been shown for *Delonix regia* dyed silk fabrics. The samples showed better dye uptake than control (untreated) as shown by increase in K/S value and color strength measured by the color measurement of dyed fabric using color scan machine. It was found that in one step dyeing process lipase was the best option. The order of reactivity of enzymes in one step process observed was Lipase > Diasterase > Protease-amylase=pyrus (biomordant) as shown in Table 4B.6. Similarly for two step dyeing process Protease and amylase combination enzymes was the best option. The order of reactivity of enzymes in one step process was observed to be Protease-amylase Lipase > *Pyrus* (biomordant) > Diasterase. Table 4B.6 shows the colorimetric values of dyed silk fabric with *Delonix* after pretreatment with different enzymes and a biomordant, the one step and two step dyeing with different pretreatments imparted a shade change from light pink to dark brown. Varied hues of color can be obtained by premordanting the silk, the different enzymes and biomordant not only cause difference in hue color but also produce significant change in K/S values. CIE L a* and b* values also show small change due to enzyme treatment. The best values are obtained for Protease + amylase and biomordant in two step process. Overall it can be concluded that in the results of enzymic and biomordant treatments, the two step process was better in terms of larger K/S values, color coordinate values, and dye adherence ability.

Table 4B.6 One step and two step dyeing processes with enzymes and biomordants

	L	a*	b*	C	H	K/S	%Age
First step control	62.157	7.919	5.109	9.424	32.815	15.174	100.00
Diasterase	64.684	6.212	13.060	14.462	64.536	24.710	162.84
Lipase	64.337	4.492	12.441	13.227	70.692	25.088	165.34
P & A	64.332	6.500	13.743	15.287	64.153	22.865	150.69
Biomordant	64.685	6.500	13.743	15.287	64.153	22.865	150.69
Second step control	62.157	7.919	5.109	9.424	32.815	15.174	100.00
Diasterase	50.831	11.241	23.251	25.826	64.172	36.564	75.87
Lipase	50.570	10.896	22.869	25.332	64.498	50.791	105.24
P & A	50.596	10.482	22.878	25.165	65.358	51.127	106.09
Biomordant	51.001	10.921	23.761	26.151	65.289	47.590	98.75

4B.3.3 Evaluation of metal mordant treated dyed Silk for change in K/S and color strength

Metal mordant treatments were also carried out with compounds such as Alum, Copper sulfate, and ferrous sulfate as shown in Table 4B.7. The purpose of using metal mordants was to compare the results with enzyme and biomordanted treated fabric swatches. It was found that the results were comparable with that of enzyme and biomordant. In the metal mordanting process we carried out one step and two step processes and it was observed that the results of one step were far better than the two step method. The mordant activity of the three mordants sequences was as follows: For one step dyeing process the order of reactivity was $Cu \rightarrow Al \rightarrow Fe$ and for two step dyeing process $Fe \rightarrow Cu \rightarrow Al$, The dyed fabric swatches are shown in the Table 4B.7. This also explains the better reactivity of biomordant as it contains copper metal in fruit extract.

Table 4B.7 shows the colorimetric values of dyed silk fabric with *Delonix* after pretreatment with different metal mordants, the dyeing with different mordants imparted a shade change from pinkish brown to dark brown. Varied hues of color were obtained from premordanting the silk with FeSO4, $CuSO_4$, and alum using aqueous extract of *Delonix* flower as shown in the Tables 4B.8 and 4B.9, the different mordants not only cause difference in hue color and significant changes in K/S values but also L values and brightness index values. The best values were obtained with ferrous sulfate for two steps and copper sulfate for one step dyeing. This was for the first time that simultaneous mordanting-dyeing results were better than step wise dyeing process in the case of metal mordanting.

Tables 4B.8 and 4B.9 show the fastness properties of one and two step dyeing process of dyed silk swatches using biomodant and enzymes. The order of reactivity of enzymes in one step process was found to be Lipase > Diasterase > Protease-Amylase = *Pyrus* (biomordant). Similarly for two step dyeing process Protease and amylase combined enzyme was the best option. The order of reactivity of enzymes in two step process observed was Protease-Amylase > Lipase > Pyrus (biomordant) > Diasterase. This clearly showed that Lipase in one step and Protease-amylase in two steps are the choice of enzyme for *Delonix* flower extract. Another experiment was carried out where the enzyme treatment to the silk fabric was done by raising the temperature of the dye bath from 30°C to 70°C for the purpose of denaturing the enzyme (Vankar and Shanker, 2009). Poor dyeability was observed in this case. Thus the role of enzyme in both the cases has been demonstrated to fix the dye molecules on the fiber surface which was not observed in the case of control sample (devoid of enzyme treatment) or in the case where denatured enzyme was used. The enzymes are absorbed by virtue of various ionic and nonionic forces of attraction on to the silk fabric through hydrogen bonding, dipole-dipole interactions, and electrostatic forces. The enzyme-dye complex thus formed on the surface of the dyed silk fabric acts as a barrier for not letting the dye get washed off. The probable reason for the superiority of two step dyeing process over the one step process could be due to the fact that sequential treatment to the silk fabric may be resulting in formation of the large molecular size and low water solubility complex on the surface of the silk fabric. Similarly, the probable mode of chelation of copper present in P*yrus* (biomordant) with quercetin has been proposed as discussed (Chapter 2). The metal-dye complex formation helped in better dye adherence which was not observed in the case of control sample (devoid of mordant treatment).

Table 4B.7   **One step and two step processes with metal mordants**

	L	a*	b*	C	H	K/S	%Age
First stepcontrol	62.157	7.919	5.109	9.424	32.815	15.174	100.00
Alum	63.804	−0.264	9.270	91.274	91.667	26.840	176.88
CuSO$_4$	64.936	−0.384	14.106	14.111	91.615	44.256	291.66
FeSO$_4$	64.411	0.355	13.294	13.299	88.435	24.200	159.48
Second step control	62.157	7.919	5.109	9.424	32.815	15.174	100.00
Alum	62.981	3.280	7.367	8.064	65.973	25.174	165.90
CuSO$_4$	63.624	3.402	10.463	11.002	71.959	26.820	183.33
FeSO$_4$	62.304	4.701	4.701	5.373	118.98	70.623	465.41

Table 4B.8   **Fastness properties of dyed silk fabrics with biomordant and enzyme by one step process**

Dyeing methods	Wash-perspiration-rubbing-light					
	WF	Per$_{acidic}$	Per$_{basic}$	Rub$_{dry}$	Rub$_{wet}$	LF
Control	1–2	1–2	1–2	1	1	1
Biomordant	2	2	2	2–3	2–3	2
Diasterase	2–3	2–3	2–3	2–3	3	2–3
Protease-amylase	2–3	2–3	2–3	2–3	2	2
Lipase	3	3	3	3	3	3

Table 4B.9   **Fastness properties of dyed silk fabrics with biomordant and enzyme by two step process**

Dyeing methods	Wash-perspiration-rubbing-light					
	WF	Per$_{acidic}$	Per$_{basic}$	Rub$_{dry}$	Rub$_{wet}$	LF
Control	2	2	2	1–2	1–2	1
Biomordant	3	3–4	3–4	3	3	3
Diasterase	2–3	2–3	2–3	2–3	3	2–3
Protease-amylase	4	4	4	4	4	4
Lipase	3–4	3–4	3–4	3–4	3–4	3

This is probably the first report where *Delonix* flower extract has been shown as a source of natural dye used in conjunction with enzymes and biomordant and gave very subtle shades of pink (Fig. 4B.3). The above experiments showed that enzymatic and biomordant treatment can give good color strength to silk fabric using *Delonix* (Gulmohar) flower as a dye source and has good potential for commercial dyeing. It is a nontoxic dye. Use of enzyme and biomordant were deliberate attempt to avoid metal mordanting in silk dyeing as it would make textile dyeing more ecofriendly. The order of reactivity of enzymes in one step process was found to be Lipase > Diasterase > Protease-Amylase = *Pyrus* (biomordant). Similarly for two step dyeing process the order of reactivity of enzymes in two step process observed was Protease-Amylase > Lipase > Pyrus (biomordant) > Diasterase. Protease and amylase combination enzyme was the best option.

Overall it can be concluded that in the case of enzymic treatment and biomordant the two step process was better in terms of larger K/S values, color coordinate values, and dye adherence. It was shown that the former two give comparable results with metal mordanted samples and thus were suited for industrial silk dyeing and the range of colors are good too.

Fig. 4B.3 Shade card dyed silk by *Delonix regia.*

4B.4   Silk dyeing with *Plumeria rubra*

4B.4.1   Dyeing

Plumeria are beautiful pink flower easily available and their dyeing potential on different fabric can be ascertained (Rupali and Alka, 2014). The silk was dyed with dye extract (Vankar and Shankar, 2007) keeping M:L ratio as 1:40, in case of silk and the pH was maintained at 4 by adding buffer solution (sodium acetate and acetic acid). Dyeing was done by conventional dyeing method as well as by sonicator. In each case the dyed material was washed with cold water and dried at room temperature, it was then dipped in brine for dye fixing. The color strength was determined calorimetrically using Colorscan at the maximum wavelength of the natural colorant. The dye uptake by silk fabric is shown in Fig. 4B.4.

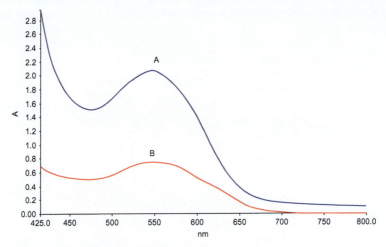

Fig. 4B.4 Dye uptake using fresh pink *Plumeria* flower extract.

4B.4.2 Sonicator dyeing

Generally, the sonochemical activity arises mainly from acoustic cavitation in liquid media. The acoustic cavitation occurring near a solid surface will generate microjets the microjet effect facilitates the liquid to move with a higher velocity resulting in increased diffusion of solute inside the pores of the fabric. In the case of sonication, localized temperature raise and swelling effects due to ultrasound may also improve the diffusion.

The stable cavitation bubbles oscillate which is responsible for the enhanced molecular motion and stirring effect of ultrasound. In case of cotton dyeing, the effects produced due to stable cavitation may be realized at the interface of leather and dye solution. Dye uptake was studied during the course of the dyeing process for a total dyeing time of 1 h with and without ultrasound. Dye uptake showed 81% and 67% respectively. Fig. 4B.5 showed the dye uptake by sonicator and conventional dyeing method respectively.

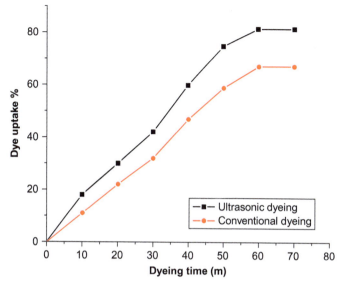

Fig. 4B.5 Dye uptake by sonicator and conventional dyeing methods.

4B.4.3   Optimization of mordants with K/S and color hue changes

Different mordants are used in 2%–4% keeping in mind the toxicity factor of some mordants. Varied hues of color can be obtained from premordanting the silk fabric with $FeSO_4$, $SnCl_2$, $CuSO_4$, $SnCl_4$, $K_2Cr_2O_7$, and alum and were dyed by aqueous extract of *Plumeria* as shown in the Table 4B.10 with different mordants that not only cause difference in hue color and significant changes in K/S values but also L values and brightness index values. This strong coordination tendency of Al enhances the interaction between the fabric and the dye, resulting in high dye uptake and brighter shade, while all other metals show less coordination and lighter shades. This is clearly shown in Fig. 4B.6.

It was observed that dyeing with *Plumeria* gave fair to good fastness properties in conventional dyeing for 1 h showing good dye uptake. The structure of both the colorants, rutin and quercitrin molecules show the presence of 3–5 hydroxyl groups which rightly suited for metal chelation. The Table 4B.10 shows L, a*, and b* values and fastness properties for different mordants for silk, some show higher value of L showing lighter shades while lower L values signify deeper shades. The colorfastness to washing was between 5 and 4–5, for silk. The results clearly show that sonicator dyeing is better in terms of better dye uptake, reduced dyeing time and cost effectiveness. Overall, it could be used for commercial purpose; the dyed silk fabrics attain acceptable range of colors as depicted in shade card (Fig. 4B.7).

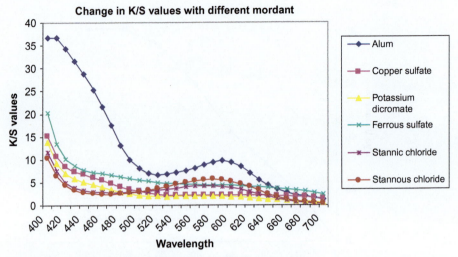

Fig. 4B.6 Showing change in K/S values with different mordants.

Fig. 4B.7 Shade card of silk dyed by *Plumeria* flower.

Table 4B.10 **Color obtained, fastness properties and L*, a*, and b* values for dyed silk with pink plumeria at λ_{max} 421 nm**

Mordant	Color obtained	Wash fastness IS-687-79	Light fastness IS-2454-85	Rubbing fastness IS-766-88	L*	a*	b*
Alum	Bright green	5	V	4–5	52.15	−10.5	25.48
Ferrous sulfate	Steel grey	5	V	4	42.19	−2.37	3.54
Stannous chloride	Dull mauve	4/5	IV	4	50.54	−3.05	11.67
Copper sulfate	Light green	4–5	V	4	49.62	−8.11	23.57
Potassium dichromate	Light green	4–5	V	4	49.80	−4.40	22.01
Stannic chloride	Dull mauve	4/5	IV	4	52.84	−2.87	18.67

Aqueous extract of pink *Plumeria* flowers give light green to dark green to mauve shades to silk and wool with good fastness properties both by conventional as well as by sonicator dyeing. It is envisaged that bright green shades in case of alum mordant in silk, is due to aluminum ion binding to the 3-hydroxychromane groups. The mordants used are in 2%–4% only. Although the dye extract is red in color and varies slightly with change in pH, the dye still has a good scope in the commercial dyeing of silk fabric for garment industry.

4B.5 Dyeing with Combretum (*Quisqualis indica*) flowers

4B.5.1 Preparation and optimization of aqueous extract of Combretum indicum

The flowers of *Combretum indicum* were found to give out color in hot water very easily. The flowers were frozen after collection and then dipped in hot boiling water to get the maximum color in 30 min which showed deepening of hue color. Increasing the quantity of flowers from 2 to 20 g per 100 mL water boiled for 60 min is accompanied with an increase in color strength and depth in color hue (Srivastava and Vankar, 2011).

4B.5.2 Optimization of mordants with K/S and color hue changes

Although researchers (Sati et al., 2003) have dyed wool, attempts have been carried out to dye cotton and silk with the *Combretum* flower extracts. Different mordants were used in 2%–4%, keeping in mind the toxicity factor of some mordants. Varied hues of color were obtained from premordanting the cotton and silk fabrics with $FeSO_4$, $K_2Cr_2O_7$, and alum and were dyed by aqueous extract of *Combretum indicum*. Commonly, cellulosic fibers have to be premordanted with tannic acid that gives carboxylic acid groups (−COOH) to the fibers. Then the mordant provides the necessary chelation for the dye molecules. As shown in Fig. 4B.8, the different mordants not only cause difference in hue color and significant changes in K/S values but also L values and brightness index values.

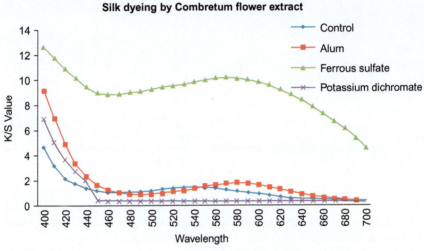

Fig. 4B.8 K/S values for dyed silk fabrics with *Combretum indicum*.

It was observed that dyeing with *Combretum indicum* flower extract gave fair to good fastness properties in conventional dyeing. Dyeing for just 1 h showed good dye uptake. Table 4B.11 shows L, a*, and b* values and can be seen that mordants which show higher values of L show lighter shades while lower L values signify deeper shades by conventional dyeing method. Similarly negative a* and negative b* represents green and blue respectively for cotton and silk fabrics. The results clearly show that dyeing with the extract of *Combretum indicum* flower extract shows good dye uptake, in just 1 h. This process showed reduced dyeing time and cost affectivity, as the flower is available in abundance during the flowering time. Overall, it could be used for commercial purpose; the dyed fabrics attain acceptable range in terms of fastness properties as shown through K/S values in Fig. 4B.8 for silk fabric.

It was observed that dyeing with *Combretum indicum* gave good fastness properties in conventional dyeing. The highest K/S value was obtained for the ferrous sulfate in the case of silk. The CIE Lab values for cotton and silk samples are shown in Table 4B.11. Analysis of the data of table I reveals that the results of CIE Lab values and K/S values are consistent. The modest values of L in the case of alum in silk also indicates that the coloristic efficiency or tinctorial value of this natural dye for these fabrics is not high, however it is within acceptable limits.

Table 4B.11 L*, a*, and b* values for dyed silk with *Combretum indicum*

Mordant (premordanting)	Color obtained	L*	a*	b*	C	H	K/S
Control	Mauve	57.05	9.61	3.28	10.15	18.86	20.99
Alum	Greenish grey	60.15	−7.13	5.52	9.02	142.23	31.18
Ferrous sulfate	Dark steel grey	56.03	0.59	−0.06	0.91	310.41	167.83
Potassium dichromate	Grey	58.45	1.26	5.26	5.40	76.48	28.06

Table 4B.11 shows the colorimetric values of dyed silk fabric with *Combretum indicum* after pretreatment with different metal mordants, the dyeing with different mordants imparted a shade change from mauve to grey. Different mordants are used in 1%–2% keeping in mind the toxicity factor of some mordants. Varied hues of color can be obtained from premordanting the silk with $FeSO_4$, $K_2Cr_2O_7$ and alum and silk was dyed by aqueous extract of *Combretum indicum* flower, the different mordants not only cause difference in hue color and significant changes in K/S values but also L values and brightness index values. The best values are obtained with ferrous sulfate. Table 4B.12 shows the results of fastness properties for unmordated dyed with some exemplary metal mordanted cotton and silk dyed fabrics.

Table 4B.12 Fastness properties of dyed silk fabrics with different metal mordants

Dyeing methods	Wash-perspiration-rubbing-light					
	WF[a]	Per$_{acidic}$	Per$_{basic}$	Rub$_{dry}$	Rub$_{wet}$	LF[b]
Silk (control)	2	2	2–3	2–3	2	2
Silk (Alum)	4	4	4	4	4	4
Silk (Fe SO$_4$)	5	4–5	4–5	4–5	4–5	4–5
Silk ($K_2Cr_2O_7$)	4–5	4	4	4	4	4–5

[a] Wash fastness.
[b] Light fastness.

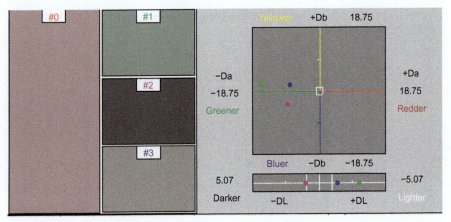

Fig. 4B.9 Different color obtained by different mordants on silk.

Fig. 4B.9 shows the colorimetric values of dyed silk fabric with *Combretum indicum* after pretreatment with different metal mordants (#0- in the order of unmordanted, alum, ferrous sulfate and potassium dichromate,). The dyeing with different mordants imparted a shade change from shades of mauve to steel grey. The lightness value decreased for iron, dichromate mordanted silk samples and shade of depth became dull and dark, while the highest was obtained with ferrous sulfate and potassium dichromate having good brightness. Aqueous extract of *Combretum indicum* flowers yield brown to green shades on silk fabric with good fastness properties. The color strengths (K/S values) are good particularly for silk mordanted by copper sulfate. The dye has good scope in the commercial dyeing of silk fabric for garment industry.

4B.6   Silk dyeing with *Ixora coccinea*

Silk was dyed with dye extract, keeping M:L ratio as 1:30 and the pH was maintained at 4 by adding buffer solution (sodium acetate and acetic acid). Dyeing was done by conventional dyeing method as well as by sonicator. In each case the dyed material was washed with cold water and dried at room temperature, it was then dipped in brine for dye fixing. The color strength was determined colorimetrically using Colorscan at the maximum wavelength of the natural colorant (Vankar and Shanker, 2006a,b).

Dye uptake was studied during the course of the dyeing process for a total dyeing time of 1 h with and without ultrasound. Dye uptake showed 84% and 43%, respectively as shown in Fig. 4B.10.

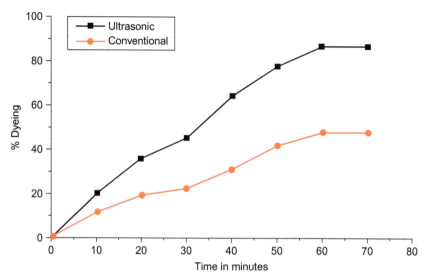

Fig. 4B.10 Dye-uptake by different methods for *Ixora*.

4B.6.1   Optimization of mordants with K/S and color hue changes

Different mordants are used in 2%–4% keeping in mind the toxicity factor of some mordants. Varied hues of color can be obtained from premordanting the silk fabrics with $FeSO_4$, $SnCl_2$, $CuSO_4$, $SnCl_4$, $K_2Cr_2O_7$ and alum were dyed by aqueous extract of *Ixora* as shown in the Table 4B.13; the different mordants not only cause difference in hue color and significant changes in K/S values but also L values and brightness index values. Copper and Iron exhibited the highest K/S values, due to their ability to form coordination complexes with the dye molecules. This strong coordination tendency of iron enhances the interaction between the fabric and the dye, resulting in high dye uptake and shows darker shades, while all other metals show some extent of coordination. This is clearly seen in the Fig. 4B.11.

It was observed that dyeing with red *Ixora* flowers gave fair to good fastness properties in conventional dyeing for 1 h showed good dye uptake. The structure of the dye molecules show that there is presence of 3–5 hydroxyl groups which makes them very good substrates for metal chelation. The Table 4B.13 shows L, a*, and b* values and fastness properties for different mordants for silk, higher value of L shows lighter shades while lower L values signify deeper shades. Similarly negative a* and negative b* represents green and blue, respectively. The colorfastness to washing was between 4 to 4–5 to 5, for silk. The results clearly show that sonicator dyeing is better in terms of better dye uptake, reduced dyeing time and cost effectiveness. Shades ranging from purple, mauve, and to earthy brown are obtained from *Ixora* (Fig. 4B.12).

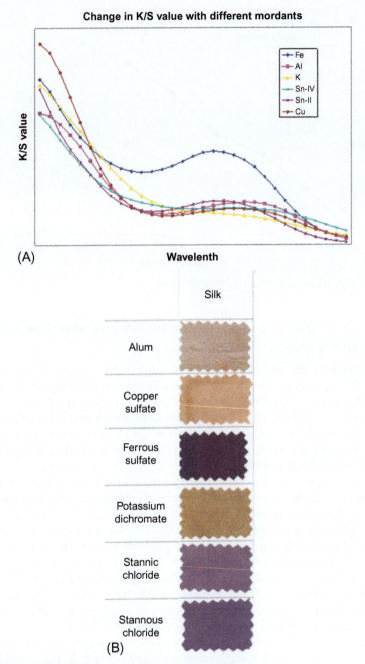

Fig. 4B.11 Change in K/S values with different mordants.

Table 4B.13 Color obtained, fastness properties and L*, a*, and b* values for dyed silk with ixora flowers at λ_{max} 421 nm

Mordant	Color obtained	Wash fastness IS-687-79	Light fastness IS-2454-85	Rubbing fastness IS-766-88	L*	a*	b*
Alum	Mauve	5	V	4–5	48.88	2.39	−3.71
Ferrous sulfate	Black	5	V	4	49.65	−2.11	0.95
Stannous chloride	Dark purple	4/5	IV	4	49.26	6.07	−12.92
Copper sulfate	Skin brown	4–5	V	4	53.09	5.02	13.72
Potassium dichromate	Greenish grey	4–5	V	4	53.08	−2.42	11.21
Stannic chloride	Dark purple	4/5	IV	4	50.12	5.17	10.08

Aqueous extract of red *Ixora* flowers give dark green to purple shades on silk with good fastness properties both by conventional as well as by sonicator dyeing. The mordants used are in 2%–4% only. Although the dye extract is red in color and varies slightly with change in pH, the dye gives shades of green and purple mainly however, in the case of silk dyeing on Fe mordanted fabric it gives black shade. The dye has good scope in the commercial dyeing of silk for garment dyeing for both domestic and international market.

4B.7   Silk dyeing with *Bischofia javanica Bl.* (Maub)

4B.7.1   Preparation of Munga silk

The munga silk fabric was scoured with solution containing 0.5 g/L sodium carbonate and 2 g/L nonionic detergent (Labolene) solution at 40–45°C for 30 min, keeping the material to liquor ratio at 1:50. The scoured material was thoroughly washed with tap water and dried at room temperature. The scoured material was soaked in clean water for 30 min prior to dyeing or mordanting.

4B.7.2   Dyeing

Dyeing was carried out in the following manner: A two step dyeing (in the ratio of 2% mordant, owf) was used as pretreatment and then dyeing with *Bischofia* extract

(10%, owf) was carried out for 1 h at temperature (Vankar et al., 2007a), in sonicator bath having temperature of 30–40°C. A similar experiment was carried out for 1 h by conventional dyeing for the sake of comparison. The dyed fabrics were dipped in dye-fix (2%, 15 min) and then rinsed thoroughly in tap water and allowed to dry in open air (Table 4B.14).

Dye uptake was studied during the course of the dyeing process for a total dyeing time of 1–3 h with and without ultrasound. About 58% exhaustion of dye (*Bischofia*) could be achieved in 1 h dyeing time using ultrasound while only 40%, in the absence of ultrasound in stationary condition was observed in the case of alum mordant as shown in Fig. 4B.13. Similarly dye uptake was calculated for other mordants with and without sonication in order to show the beneficial effect of sonication in natural dyeing. The CIE Lab values for silk dyed swatches shown in Table 4B.15. K/S were measured for silk fabrics as shown in Fig. 4B.14.

4B.7.3   Effect of mordanting conditions

It was observed that the premordanting technique with metal mordants imparted good fastness properties to the silk fabric. Therefore, in premordanting technique, the dyed fabrics were mordanted with stannic chloride, stannous chloride, ferrous sulfate, copper sulfate, potassium dichromate, and alum. The absorption of color by silk fabric was enhanced when using metal mordants, this might be due to the maximum absorption and easy formation of metal-complexes with the fabric.

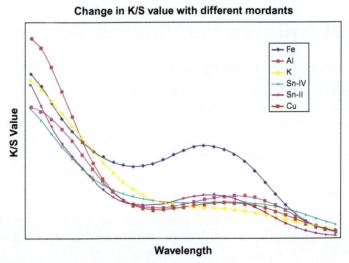

Fig. 4B.12 Change in K/S value with different mordants.

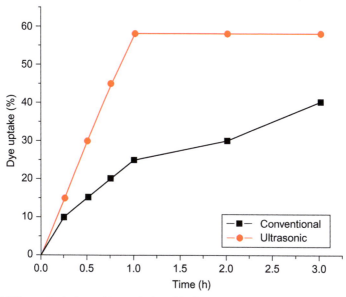

Fig. 4B.13 Ultrasonic dyeing with *Bischofia* with alum mordant.

Table 4B.14 **L*, a*, b*, C, and H values for Silk Fabric dyed with** *Bischofia javanica* **(Maub)**

Method	Mordant	L*	a*	b*	C	H
Premordanting	Alum	65.802	−1.104	26.251	26.274	92.443
	Copper sulfate	60.609	−0.460	27.873	27.877	90.981
	Stannous chloride	65.196	−1.458	25.411	25.453	93.319
	Ferrous sulfate	23.445	0.790	1.112	1.364	54.587
	Pot. dichromate	60.954	−0.082	24.318	24.318	90.229
	Stannic Chloride	65.055	−0.728	23.233	23.244	91.830

Table 4B.15 **Fastness properties of dyed silk fabrics under conventional heating and ultrasonic conditions of metal modanting and** *Bischofia*

Dyeing methods	Wash-perspiration-rubbing-light					
	WF	Per$_{acidic}$	Per$_{basic}$	Rub$_{dry}$	Rub$_{wet}$	LF
Conventional	4	4	4	4	4	4
Ultrasonic	5	5	5	5	5	6

WF = wash fastness, LF = light fastness.

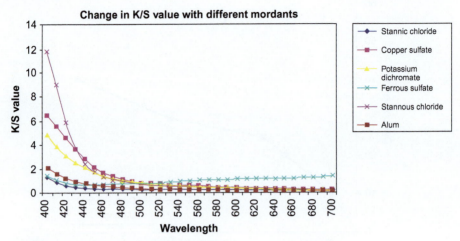

Fig. 4B.14 Change in K/S values with different mordants for silk fabrics.

Fig. 4B.15 Shade card of silk dyeing with *Bischofia*.

The dye uptake in case of two steps dyeing is 50%, 45%, and 58% with mordants—Fe, Cu, and alum mordant by sonicator method. The effectiveness of metal mordant-*Bischofia* in better dye uptake appears to be an improved process resulting in good fastness properties and thus sonication outweighs conventional dyeing method.

Bischofia javanica was found to have good agronomic potential as a dye crop in Arunachal Pradesh. Metal mordant when used in conjunction with *Bischofia javanica* was found to enhance the dyeability and fastness properties. The dye uptake in case of two steps dyeing is 50%, 45%, and 58% with mordants—Fe, Cu, and alum mordant by sonicator method. Thus the net enhancement of dye uptake was 22%, 25%, and 18% for mordants—Fe, Cu, and alum, respectively and shades of yellow to brown were obtained as shown in shade card (Fig. 4B.15).

Conclusion

India has the unique distinction of being bestowed by nature with many sources of dyes still unexplored. In this respect silk is a favorable fabric and easy to work with new natural dyes. The tedious application process and nonreproducibility of shades and insufficient availability are some of the factors responsible for their nonadoption in mainstream textile processing. However, at the present level of dye resource availability, their adoption by the textile industry is not desirable also as that would result in an environmental disaster by way of loss of biodiversity and depletion of forest cover in spite of the tremendous environmental advantage offered by them in terms of the lower pollution of the effluent if used properly. This advantage can be utilized by the traditional artisans in preserving their surroundings from the ill effects of pollution caused by synthetic dyes as they do not have access to expensive effluent treatment plants needed for synthetic dyes (Saxena and Raja, 2014). A clean production form offered by natural dyes is a better option for them. The proceeds of research conducted on the development of improved application procedures for better fastness and environmental agreement should reach people so that they can earn their living and the buyer also get the benefit of truly environmental friendly textiles.

Further Reading

- P.S., Vankar, 2013. *Handbook on Natural Dyes for Industrial Applications (with Color Photographs).* National Institute of Industrial Research. New Delhi-110007, India
- R. Shanker, 2011. Studies on Newer Natural Dyes using Innovative Methods of Mordanting-Enzymes and Bio Mordants. Thesis submitted to CSJM University, Kanpur India, 160–170.
- H. Das, & D. Kalita, 2016. Fibers and Dye Yielding Plants of North East India. In: Purkayastha, J. (ed.) Bioprospecting of Indigenous Bioresources of North-East India. Singapore: Springer Singapore.
- X.L. He, X.L. Li, Y.P. Lv, and Q. He, 2015. Composition and color stability of anthocyanin-based extract from purple sweet potato. Food Science and Technology (Campinas), 35(3), pp. 468–473.
- K.S. Suslick, and D.J. Flannigan, 2008. Inside a collapsing bubble: sonoluminescence and the conditions during cavitation. Annu. Rev. Phys. Chem., 59, pp. 659–683.
- Katti, M.R., Kaur, R. and Shrihari, N., 1996. Dyeing of silk with mixture of natural dyes. Colourage, 43(12), pp. 37+–2.

References

Brown, E.J., Khodr, H., Hider, C.R., Rice-Evans, C.A., 1998. Structural dependence of flavonoid interactions with Cu2+ ions: implications for their antioxidant properties. Biochem. J. 330, 1173–1178.

Delgado-Vargas, F., Jiménez, A., Paredes-López, O., 2000. Natural pigments: carotenoids, anthocyanins, and betalains—characteristics, biosynthesis, processing, and stability. Crit. Rev. Food Sci. Nutr. 40, 173–289.

Mira, L., Tereza Fernandez, M., Santos, M., Rocha, R., Helena Florêncio, M., Jennings, K.R., 2002. Interactions of flavonoids with iron and copper ions: a mechanism for their antioxidant activity. Free Radic. Res. 36, 1199–1208.

Rupali, D., Alka, C., 2014. A natural colorant for silk fiber: *Plumeria rubra*. Int. J. of Life Sciences, Special Issue A2, October 2014, pp. 166–168.

Sati, O.P., Rawat, U., Sati, S.C., Srivastav, B., 2003. Optimization of procedure for dyeing of wool with *Combretum indicum* arboretum a source of natural dye. Colourage 50, 43–44.

Saxena, S., Raja, A., 2014. Natural dyes: sources, chemistry, application and sustainability issues. In: Roadmap to sustainable textiles and clothing. Springer.

Shanker, R., 2011. Studies on newer natural dyes using innovative methods of mordanting-enzymes and bio mordants. Thesis submitted to CSJM University, Kanpur, India. pp. 160–170.

Shukla, D., Vankar, P.S., 2013. Natural dyeing with black carrot: new source for newer shades on silk. J. Nat. Fibers 10, 207–218.

Srivastava, J., Vankar, P.S., 2011. Natural dyeing with dark pink flowers of *Quisqualis indica*. Asian Dyers 58, 61–63.

Vankar, P.S., Shanker, R., 2007. Dyeing silk and wool with Plumeria (pink) flower. Asian Text. J., 104–107.

Vankar, P., Shanker, R., 2006a. Sonicator dyeing of cotton and silk fabric by Ixora Coccinea. Asian Text. J., 77–80.

Vankar, P.S., Shanker, R., 2006b. Dyeing cotton, silk and wool with *Cayratia carnosa* Gagn.or Vitis trifolia. Asian Text. J. 38.

Vankar, P.S., Shanker, R., 2008. Ecofriendly ultrasonic natural dyeing of cotton fabric with enzyme pretreatments. Desalination 230, 62–69.

Vankar, P.S., Shanker, R., 2009. Potential of *Delonix regia* as new crop for natural dyes for silk dyeing. Color. Technol. 125, 155–160.

Vankar, P.S., Shukla, D., 2011. Natural dyeing with anthocyanins from *Hibiscus rosa sinensis* flowers. J. Appl. Polym. Sci. 122, 3361–3368.

Vankar, P.S., Shanker, R., Dixit, S., Mahanta, D., Tiwari, S.C., 2007a. Characterisation of the colorants from leaves of *Bischofia javanica*. Int. Dyer 192, 31–33.

Vankar, P.S., Shanker, R., Verma, A., 2007b. Enzymatic natural dyeing of cotton and silk fabrics without metal mordants. J. Clean. Prod. 15, 1441–1450.

Welch, C.R., Wu, Q., Simon, J.E., 2008. Recent advances in anthocyanin analysis and characterization. Curr. Anal. Chem. 4, 75–101.

Dyeing application of newer natural dyes on wool with fastness properties, CIE lab values and shade card

D. Shukla, P.S. Vankar
FEAT (Facility for Ecological and Analytical Testing), Kanpur Kalyanpur, India

Introduction

Wool fiber is a polypeptide and has many active sites in which a dye can bind to the fiber. It may form covalent or ionic bonding to the −NH2 and −COOH groups on the ends of the polymer or form similar interactions with the amino acid side chains resulting in significant colorimetric and fastness properties. The chemistry of bonding of dyes to fiber is very complex. In fact, bonding occurs in different ways such as direct bonding (covalent), H-bonding, and hydrophobic interactions (Van der Waals forces). Mordants help in binding of dyes to fiber by forming a chemical bridge between the dye and the fiber, thus improving the substantivity of a dye along with increase in fabric's color fastness properties and color depth as well. The effect of dye concentration on color strength (K/S) of woolen yarn dyed with Natural dye extracts such as Celosia, Nerium, Hollyhock, Hibiscus mutabilis, Caryatia, Tegetus, Rambutan, and Curcuma show very good results. Increasing the concentration of dye decreased lightness (L) values of woolen yarn samples, indicating darker shades. Different metal salts such as alum, ferrous sulfate, and stannous chloride were used to enhance the fastness properties (light, wash, dry, and wet rubs) of dyed woolen yarn. Pretreatment of woolen yarn samples with safe chemicals and metal salts has shown encouraging results with better fastness properties and enhanced color strength values. Many other aides were used in wool dyeing (Riva et al., 1999).

Wool dyeing

Commercially bleached wool yarn supplied by Jaypee (pure new wool) was used for dyeing.

Scouring of Wool: The Wool yarn was put in a bath containing 0.5 g/L sodium carbonate and 2 g/L nonionic detergent (Labolene) solution at 40–45°C for 30 min, keeping the material to liquor ratio at 1:50. The scoured wool was thoroughly washed with tap water and dried at room temperature. The scoured wool yarn was soaked in clean water for 30 min prior to dyeing or mordanting.

Natural Dyes for Textiles. http://dx.doi.org/10.1016/B978-0-08-101274-1.00011-2

Mordanting: Natural dyes require chemicals in the form of metal salts to produce an affinity between the wool yarn and the pigments and these chemicals are known as mordants. Accurately weighed wool sample was treated with different metal salts, only premordanting with metal salts was carried out before dyeing.

Dyeing: The wool yarn was dyed with dye extract, keeping M:L ratio as 1:30 and the pH was maintained at 4 by adding buffer solution (sodium acetate and acetic acid). Temperature of the dye bath was raised to 60°C over half an hour and left at that temperature for another 30 min. The dyed wool yarn was then rinsed with water thoroughly, squeezed and dried. The dyed yarn was then dipped in brine for dye fixing.

4C.1 Wool dyeing with *Celosia cristata* flower

Wool was dyed with *Celosia cristata* flower extract after Scouring, pretreatment, and mordanting as general practice.

It is for the first time that *Celosia* flowers have been used as a dye source for wool dyeing. Results of the preliminary tests carried out to standardize the dye extraction and dyeing procedures with the *Celosia* flowers showed that ethylene diamine pretreated wool showed darker shades as compared to morpholine and sodium hydroxide as seen in Tables 4C.1–4C.3. The shades obtained were brighter as well. Pretreatments were designed based on the chemical structure of the colorants. The acidic functionality in the colorant thus helps by basic pretreatment in better dye uptake and color adherence. The optimum dyeing time in dye bath was 2 h. Most of the shades obtained were shades of olive green and

Table 4C.1 **Fastness properties of the dyed wool yarn with Celosia flowers after pretreatment with ethylene diamine**

Mordant		Fastness properties					
		Wash	Light	Rubbing		Perspiration	
				Dry	Wet	Alkaline	
		IS-687-79	IS-2454-85	IS-766-88		IS-971-83	Acidic
Control	Peach	3/4	3/4	3	3	3	3
Stannic chloride	Light yellow	4	4	4	4	4	4
Stannous chloride	Yellow	4	4	4	4	4	4
Ferrous sulfate	Olive green	4	4	4	4	4	4
Alum	Bright yellow	4	4	3	3	3	3
Potassium dichromate	Greenish yellow	4	4	4	4	4	4
Copper sulfate	Dull green	4	4	4	4	4	4

Table 4C.2 Fastness properties of the dyed wool yarn with Celosia flowers after pretreatment with sodium hydroxide

Mordant		Wash IS-687-79	Light IS-2454-85	Rubbing		Perspiration	
				Dry	Wet	Alkaline IS-971-83	Acidic
				IS-766-88			
Control	Peach	3	3	3	3	3	3
Stannic chloride	Yellowish brown	4–5	4–5	4	4	4/5	4/5
Stannous chloride	Pale yellow	4–5	4–5	4	4	4	4
Ferrous sulfate	Dark greenish yellow	4–5	4–5	4	4	4	4
Alum	Yellowish brown	4	4	3	3	3	3
Potassium dichromate	Greenish yellow	4–4	4	4	4	4	4
Copper sulfate	Dull green	4–5	4–5	4	4	4	4

Table 4C.3 Fastness properties of the dyed wool yarn with Celosia flowers after pretreatment with Morpholine

Mordant		Wash IS-687-79	Light IS-2454-85	Rubbing		Perspiration	
				Dry	Wet	Alkaline IS-971-83	Acidic
				IS-766-88			
Controlled	Peach	3	3	3	3	3	3
Stannic chloride	Yellowish brown	4–5	4–5	4	4	4/5	4/5
Stannous chloride	Pale yellow	4–5	4–5	4	4	4	4
Ferrous sulfate	Dark- greenish yellow	4–5	4–5	4	4	4	4
Alum	Yellowish brown	4	4	3	3	3	3
Potassium dichromate	Greenish yellow	4–4	4	4	4	4	4
Copper sulfate	Dull green	4–5	4–5	4	4	4	4

brown but premordanting with potassium dichromate showed dark greenish yellow color. The L a* b* values of the dyed wool samples pretreated with EDA are given in Table 4C.4. Dyeing these flower in this way showed good fastness properties. After dyeing process the samples were dipped in brine and then washed with plain water. This dye can be used for dyeing in various shades of mustard/brown/green for wool. The findings of color fastness tests of wool samples to rubbing under dry conditions showed that the samples had fair to excellent fastness as compared to dry rubbing. The sample when subjected to wet rubbing exhibited a decrease in color fastness, washing fastness test showed good fastness. *Celosia* shows good prospects for wool dyeing (Shanker and Vankar, 2005a) and different shades of wool with different pretreatments (NaOH, EDA, Morpholine and only metal premordanting was represented in shade card (Figs. 4C.1–4C.4)).

Table 4C.4 **L*, a*, and b* values for dyed wool with Celosia flowers for ethylene diamine pretreated wool**

Mordant (premordanting)	Color obtained	L*	a*	b*
Alum	Bright yellow	79.083	−2.001	50.671
Stannic chloride	Light yellow	81.113	−4.472	39.843
Stannous chloride	Yellow	97.084	−1.345	40.869
Ferrous sulfate	Olive green	74.003	−6.078	28.955
Copper sulfate	Dull green	80.231	−9.377	34.909
Pot. dichromate	Dark greenish yellow	84.221	−6.033	28.794

**Cockscomb
(NaOH pretreatment)**

Stannous chloride

Stannic chloride

Alum

Copper sulfate

Ferrous sulfate

Potassium dichromate

Fig. 4C.1 Shades of the dyed wool yarn with *Celosia* flowers after pretreatment with sodium hydroxide.

**Cockscomb
(EDA pretreatment)**

Stannous chloride

Stannic chloride

Alum

Copper sulfate

Ferrous sulfate

Potassium dichromate

Fig. 4C.2 Shades of the dyed wool yarn with *Celosia* flowers after pretreatment with ethylene diamine.

**Cockscomb
(Morpholine pretreatment)**

Stannous chloride

Stannic chloride

Alum

Copper sulfate

Ferrous sulfate

Potassium dichromate

Fig. 4C.3 Shades of the dyed wool yarn with *Celosia* flowers after pretreatment with Morpholine.

Cockscomb

Stannous chloride

Stannic chloride

Alum

Copper sulfate

Ferrous sulfate

Potassium dichromate

Fig. 4C.4 Shades of the dyed wool yarn with *Celosia* flowers after pretreatment with metal mordanting.

4C.2 Wool dyeing with *Nerium oleander* flower

Wool was dyed with *Nerium oleander* flower extract after scouring, pretreatment, and mordanting as general practice.

4C.2.1 Optimization of mordants with K/S and color hue changes

Different mordants are used in 2%–4% keeping in mind the toxicity factor of some mordants. Varied hues of color can be obtained by premordanting the wool yarn with $FeSO_4$, $SnCl_2$, $CuSO_4$, $SnCl_4$, $K_2Cr_2O_7$, and alum when dyed by aqueous extract of *Nerium*. As shown in the Fig. 4C.5, the different mordants not only cause difference in hue color and significant changes in K/S values but also L* values and brightness index values.

4C.2.2 Fastness properties

It was observed that dyeing with *Nerium* gave fair to good fastness properties in sonicator dyeing. Dyeing in sonicator for 1 h showed good dye uptake. Table 4C.5 shows L, a*, and b* values and can be seen that mordants which show higher value of L

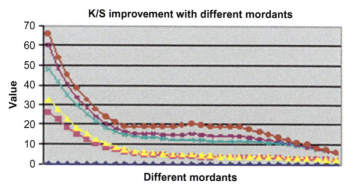

Fig. 4C.5 K/S values for wool dyed with *Nerium* at λ$_{max}$ 503.87 nm.

Table 4C.5 **Color obtained X, Y, Z and L*, a*, and b* values for dyed wool with Nerium**

Mordant (premordanting)	Color obtained	X	Y	Z	L*	a*	b*
Alum	Bright green	70.62	74.68	81.76	89.24	−0.34	−1.18
Ferrous sulfate	Dark green	5.45	5.84	5.28	29.02	−1.01	4.33
Stannous chloride	Purple	9.91	9.93	11.99	37.72	3.95	−3.68
Copper sulfate	Moss green	9.47	10.11	5.13	38.04	−0.91	20.61
Potassium dichromate	Cream	28.81	29.95	18.05	61.61	1.68	23.43
Stannic chloride	Light purple	15.16	16.39	16.09	47.48	0.48	3.21

show lighter shades while lower L values signify deeper shades. Similarly negative a* and negative b* represents green and blue respectively. The colorfastness to washing was 4–5 as shown in Table 4C.6. Dyeing with conventional method has also been compared. The results clearly show that dyeing in sonicator is better in terms of better dye uptake, reduced dyeing time, and cost effectiveness. Overall, it could be used for commercial purpose, the dyed wool yarn attains acceptable range. Fig. 4C.5 showed the K/S values improvement by using metal mordants with *Nerium* extract. The bright colors can be seen in shade card (Fig. 4C.6).

Aqueous extract of *Nerium* flowers yield cream to green to purple shades with good fastness properties as can be seen from shade card (Fig. 4C.6). The dye has good scope in the commercial dyeing of wool yarn for carpet industry (Vankar and Shanker, 2008).

Table 4C.6 **Fastness properties for wool dyed in Sonicator with Nerium**

Mordant (premordanting)	Fastness properties					
	Washing IS-687-79	Light IS-2454-85	Rubbing IS-766-88		Perspiration IS-971-83	
			Dry	Wet	Alkaline	Acidic
Alum	4–4/5	IV	4–5	4–5	4/5	4/5
Copper sulfate	4–4/5	III	3–4	3	3–4	3–4
Ferrous sulfate	4–4/5	IV	3–4	3–4	3–4	3–4
Stannous chloride	4–4/5	III	4	3–4	3/4	4
Stannic chloride	4–5	IV	4	4	4	4
Pot. dichromate	4–4/5	IV	4	4	3/4	4

Nerium

Alum

Copper sulfate

Potassium dichromate

Ferrous sulfate

Stannic chloride

Stannous chloride

Fig. 4C.6 Shade card of wool dyed by *Nerium*.

4C.3 Wool dyeing with hollyhock

4C.3.1 Sonicator dyeing

It is well documented (Ghorpade et al., 2000) that the ultrasound energy gives rise to acoustic cavitation in liquid media. Dye uptake was studied during the course of the dyeing process for a total dyeing time of 1 h with and without ultrasound. Dye uptake showed 81% and 67%, respectively as shown in Fig. 4C.7 (Vankar and Shanker, 2006).

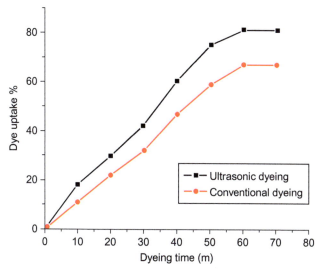

Fig. 4C.7 Comparative dye uptake by sonication and conventional method in hollyhock.

4C.3.2   Optimization of mordants with K/S and color hue changes

Different mordants are used in 2%–4% keeping in mind the toxicity factor of some mordants. Varied hues of color can be obtained by premordanting the wool yarn with $FeSO_4$, $SnCl_2$, $CuSO_4$, $SnCl_4$, $K_2Cr_2O_7$, and alum when dyed with aqueous extract of hollyhock. The different mordants not only cause difference in hue color and significant changes in K/S values but also L values and brightness index values. Copper and Iron exhibited the highest K/S values, due to their ability to form coordination complexes with the dye molecules. This strong coordination tendency of Fe enhances the interaction between the fiber and the dye, resulting in high dye uptake, while all other metals show similar coordination as shown in the Fig. 4C.8.

4C.3.3   Fastness properties of dyed wool with hollyhock

It was observed that dyeing with hollyhock gave fair to good fastness properties in conventional dyeing for 1 h and showed good dye uptake. The structure of the dye molecules shows that presence of 3–5 hydroxyl groups makes them very good substrates for metal chelation. The Table 4C.7 shows L, a*, and b* values and fastness properties for different mordants, some show higher and lower L values giving lighter and deeper shades, respectively. Similarly negative a* and negative b* represents green and blue, respectively. The colorfastness to washing was found to be between 4–5 for wool in all the cases as shown in Table 4C.7. The sonicator dyeing was compared with conventional method. The results clearly show that sonicator dyeing is better in terms of better dye uptake, reduced dyeing time, and cost effectiveness. Overall, it could be used for commercial purposes, the dyed wool yarn accomplish satisfactory range.

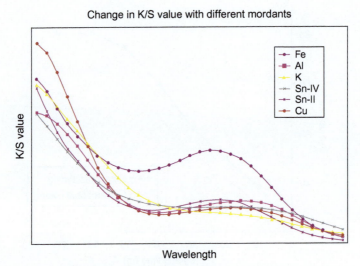

Fig. 4C.8 Change in K/S values with different mordants.

Table 4C.7 **Color obtained, fastness properties and L*, a*, and b* values for dyed wool yarn with hollyhock at λ_{max} 545 nm**

Mordant	Color obtained	Wash fastness IS-687-79	Light fastness IS-2454-85	Rubbing fastness IS-766-88	L*	a*	b*
Alum	Bright green	5	V	4–5	48.15	−9.15	17.95
Ferrous sulfate	Dark brown	5	V	4	36.67	1.84	4.84
Stannous chloride	Dull green	4/5	IV	4	47.78	−2.82	15.04
Copper sulfate	Moss green	4–5	IV	4	41.18	−3.92	18.86
Potassium dichromate	Greenish brown	4–5	IV	4	48.97	1.21	25.14
Stannic chloride	Brownish green	4/5	IV	4	47.90	−1.96	14.25

Aqueous extract of hollyhock flowers give light green to dark green shades to cotton, silk, and wool with good fastness properties both by conventional as well as by sonicator dyeing. A flavylium ring and a 3′, 4′-dihydroxy group is well suited for Fe-binding and in case of aluminum ion binding, the 3-hydroxychromane groups is envisaged. The mordants used are in 2%–4% only. Although the dye extract is red in color and varies slightly with change in pH, the dye still has a good scope in the commercial wool yarn dyeing for industry.

4C.4   Wool dyeing with *Hibiscus mutabilis*

4C.4.1   Dyeing

The wool yarn was dyed with flower extract of *Hibiscus mutabilis*. The pH was maintained at 4 by adding buffer solution (sodium acetate and acetic acid). The color strength was determined colorimetrically using Colorscan at the maximum wavelength of the natural colorant. The dye uptake by wool yarn is shown in Fig. 4C.9.

4C.4.2   K/S and color hue change and fastness properties of dyed wool

Different mordants are used in 2%–4% in premordanting of the wool yarn with $FeSO_4$, $SnCl_2$, $CuSO_4$, $SnCl_4$, $K_2Cr_2O_7$, and alum. Aqueous extract of *Hibiscus mutabilis*, the different mordants not only cause difference in hue color and significant changes in K/S values but also L values and brightness index values. It was observed that dyeing with *Hibiscus mutabilis* gave fair to good fastness properties in conventional dyeing. Table 4C.8 shows L, a*, and b* in which lighter and deeper shades gave low and high L values. Similarly negative a* and negative b* represents green and blue, respectively. The colorfastness to washing grade was 4–5 as shown in Table 4C.9 for wool. The fastness properties of dyed wool yarn were shown in Table 4C.9. Overall, it could be used for commercial purposes (Shanker and Vankar, 2007a). Wool yarn achieves pleasing range. The results clearly show that this aqueous dye is better in terms of better dye uptake, reduced dyeing time, and cost effectiveness.

Aqueous extract of Gulzuba flowers yield shades with good fastness properties. The dye has good scope in the commercial dyeing of wool yarn for industry.

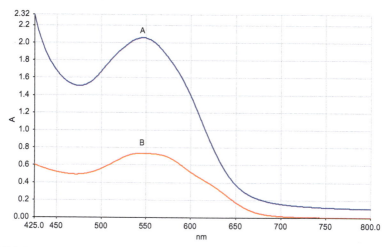

Fig. 4C.9 Dye uptake from fresh flower extract of *Hibiscus mutabilis*.

Table 4C.8 **Color obtained X, Y, Z and L*, a*, and b* values for dyed wool yarn with** *Hibiscus mutabilis*

Premordanting	Color obtained	X	Y	Z	L*	a*	b*
Alum	Bright green	70.62	74.68	81.76	89.24	−0.34	−1.18
Ferrous sulfate	Dark green	5.45	5.84	5.28	29.02	−1.01	4.33
Stannous chloride	Purple	9.91	9.93	11.99	37.72	3.95	−3.68
Copper sulfate	Moss green	9.47	10.11	5.13	38.04	−0.91	20.61
Potassium dichromate	Green	28.81	29.95	18.05	61.61	1.68	23.43
Stannic chloride	Light purple	15.16	16.39	16.09	47.48	0.48	3.21

Table 4C.9 **Fastness properties for wool yarn dyed with** *Hibiscus mutabilis*

Premordanting	Fastness properties					
	Washing IS-687-79	Light IS-2454-85	Rubbing IS-766-88		Perspiration IS-971-83	
			Dry	Wet	Alkaline	Acidic
Alum	4–4/5	IV	4–5	4–5	4/5	4/5
Copper sulfate	4–4/5	III	3–4	3	3–4	3–4
Ferrous sulfate	4–4/5	IV	3–4	3–4	3–4	3–4
Stannous chloride	4–4/5	III	4	3–4	3/4	4
Stannic chloride	4–5	IV	4	4	4	4
Pot. dichromate	4–4/5	IV	4	4	3/4	4

4C.5   Wool dyeing with *Cayratia cornosa*

4C.5.1   Dyeing

The wool yarn was dyed with the dye extract of *Cayratia cornosa* blue berries and the pH was maintained at 4–5 by adding buffer solution. Temperature of the dye bath was raised to 60°C over half an hour and left at that temperature for another 30 min. The dyed yarn was then squeezed and dried. The dyed material was then dipped in brine for dye fixing (Vankar and Shanker, 2006) (Fig. 4C.10).

4C.5.2   Fastness properties of dyed fabrics

The results are very encouraging; it is for the first time that *Cayratia* fruits/berries have been used as a dye source for textile dyeing. Results of the preliminary tests carried out to standardize the dye extraction and dyeing procedures with the *Cayratia*

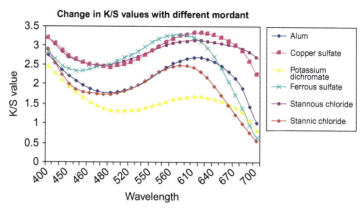

Fig. 4C.10 K/S of wool yarn dyed with *Cayratia cornosa*.

fruits showed good dyeing results in Table 4C.10 for wool. The shades obtained were bright as well as having very good fastness properties. After the dyeing process the samples were dipped in brine and then washed with plain water. This dye can be used for dyeing in various shades of blue-dark blue and purple shades for wool (shade card Fig. 4C.11). The findings of color fastness tests of all the samples to rubbing under dry conditions showed that the samples had fair to excellent fastness as compared to dry rubbing. *Cayratia* shows good prospects for textiles especially wool dyeing (Table 4C.11).

Table 4C.10 Fastness properties of the dyed wool with Cayratia

	Fastness properties					
	Wash	**Light**	**Rubbing**		**Perspiration**	
			Dry	**Wet**	**Alkaline**	
Mordant	**IS-687-79**	**IS-2454-85**	**IS-766-88**		**IS-971-83**	**Acidic**
Control	3/4	3/4	3	3	3	3
Stannic chloride	4	4	4	4	4	4
Stannous chloride	4	4	4	4	4	4
Ferrous sulfate	4	4	4	4	4	4
Alum	4	4	3	3	3	3
Potassium dichromate	4	4	4	4	4	4
Copper sulfate	4	4	4	4	4	4

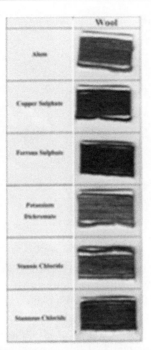

Fig. 4C.11 Shade card of wool dyed by *Cayratia carnosa*.

Table 4C.11 **L*, a*, and b* values for dyed wool with Cayratia (premordanting)**

Mordant	Color obtained	L*	a*	b*	C	H
Alum	**Dull green**	**47.044**	**−5.228**	**−2.487**	**5.789**	**05.431**
Stannic chloride	Steel grey	46.629	−2.133	−3.185	3.833	236.167
Stannous chloride	Dark purple	46.667	−2.366	−2.258	3.271	223.644
Ferrous sulfate	Olive green	46.002	0.386	−4.961	4.976	274.483
Copper sulfate	Dark green	46.785	−3.605	−2.524	4.401	214.983
Pot. dichromate	Light green	48.027	−5.257	2.958	6.032	150.646

4C.6 Wool dyeing with *Tagetes erecta*

4C.6.1 Dyeing

Dyeing was carried out in following manner: A stepwise dyeing of pretreatment of wool mordanting (in the ratio of 1% or 2% mordant, owf) was used for dyeing with aqueous extract of *Tegetes* (5%, owf) at its original pH. The dyeing process was carried out for 2 h at temperature of 30–40°C (Vankar et al., 2009) (Vankar, 2009). The dyed yarn was dipped in saturated brine solution (15 min) which acts as dye-fix and then rinsed thoroughly in tap water and the dyed wool was allowed to dry in open air. The colorimetric data obtained from dyed yarn, which had been pretreated with only metal mordants in the cases of wool reveal that pretreatment markedly improved the wash fastness, in terms of change of shade of the dyed yarn with respect to controlled samples. It also increased the color strength and flattened the shade of the dyeing. In each experiment control dyed samples were also prepared. Through an innovative extraction process, it has been exemplified that the dye contents obtained by ethanolic extraction was higher than that obtained by aqueous extraction as shown in Table 4C.12. The ethanol after each extraction was recovered and has been reused for three consecutive times. For further use the ethanol had to be dried and distilled.

Through the different extraction processes it has also been shown that the process is ecofriendly, since the ethanol is completely recovered in rotatory evaporator and reused. Dyeing shows deeper shades. The role of the metal mordanting in natural dyeing has been well demonstrated (Sarkar and Seal, 2003; Popoola, 2000). The mordant forms an insoluble complex between dye molecule and metal, which is responsible for its insolubility and resultant affinity to the fiber, mainly silk and wool (Tiedemann and Yang, 1995).

Tegetes extract showed very good affinity for proteinaseous fibers when used in conjunction with metal mordants. The hue colors developed through metal mordanting were quite dark and saturated and fitted very well into the basic criteria of good natural dye formulation. K/S was measured for wool yarn and CIE lab values are shown in Table 4C.13.

Table 4C.12 Dye content from *Tagetes* flowers by different solvents

Part of the plant from 50 g	Dye content by aqueous method	Dye content by ethanolic method	Color of the aq. extract	Color of the eth. extract
Tagetes flower	a. 2.40 g b. 2.37 g c. 2.38 g	a. 3.55 g b. 3.45 g c. 3.48 g	Yellow	Dark yellow

Table 4C.13 L*, a*, b*, C, and H values for wool yarn dyed with *Tagetes erecta*

Method	Mordant	L*	a*	b*	C	δE	K/S
Premordanting	Controlled	52.14	−2.00	43.34	43.38	–	132.50
	Alum	58.03	10.80	60.83	61.78	22.47	84.88
	Copper sulfate	37.18	2.03	23.82	23.90	24.92	67.08
	Stannous chloride	59.31	12.80	65.44	66.68	27.55	102.12
	Ferrous sulfate	32.85	2.45	13.46	13.68	35.84	256.54
	Pot. dichromate	36.90	5.55	22.96	23.62	26.54	81.18
	Stannic chloride	67.98	13.50	80.56	81.69	43.33	234.06

Table 4C.13 showed the colorimetric values of dyed wool yarn with *Tegetes* after pretreatment with different metal mordants, the dyeing with different mordants imparted a shade change from light yellow to brown. Varied hues of color can be obtained by premordanting the wool yarn with $FeSO_4$, $SnCl_2$, $CuSO_4$, $SnCl_4$, $K_2Cr_2O_7$, and alum when dyed by aqueous extract of *Tegetes* flower. As shown in the Fig. 4C.12, the different mordants not only cause difference in hue color and significant changes in K/S values but also L values and brightness index values. The best values are obtained with ferrous sulfate. Thus in all the three cases iron provided dark shades.

The colorimetric values of dyed wool yarn with *Tegetes* after pretreatment with different metal mordants (#0–6 in the order of unmordanted, alum, copper sulfate, ferrous sulfate, potassium dichromate, stannous chloride and stannic chloride) has been shown (Fig. 4C.13). The dyeing with different mordants imparted a shade change from mustard yellow to moss green, brown and black. The lightness value decreased for iron, copper, and chromium mordanted wool samples and shade of depth became dull and dark, while the highest was obtained with stannic chloride having good brightness. The shades of the dyed wool samples were the brightest.

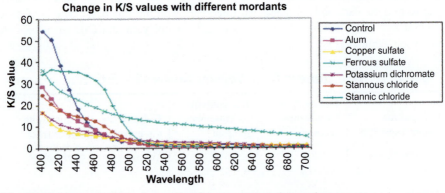

Fig. 4C.12 Change in K/S values with different mordants for wool yarn.

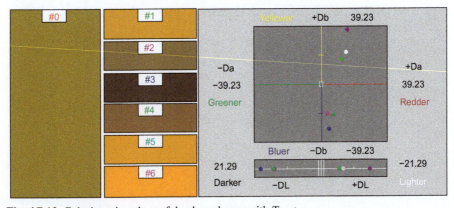

Fig. 4C.13 Colorimetric values of dyed wool yarn with *Tegetes*.

The dyed samples were tested for fastness according to Indian Standard methods and the results for alum, ferrous sulfate, and copper sulfate mordanted have been tabulated in Table 4C.14. It shows that these mordants have caused improved fastness properties in wool. Similar observations were made in the case of other three mordants as well. Marked improvement was noticed in the case of washing and light fastness. Thus the dye can be recommended for commercial use for wool yarn.

Through the UV-Visible absorption of the dyebath before and after dyeing it has been tried to show that dye uptake by mordanted fabric is better than the control samples. The enhancement in dye uptake with metal mordanting has been shown in Table 4C.15. The results clearly show marked enhancement. Overall enhancement of dye uptake due to metal mordanting has been found to be 37%–51% in wool with respect to the control samples.

Metal mordant in conjunction with *Tegetes erecta* flower extract was found to enhance the dyeability and fastness properties effect in wool. Thus the net enhancement of dye uptake due to metal mordanting has been found to range from 37%–51% in wool with respect to the control samples. The higher percentage of color strength in the case of wool makes T*egetes erecta* best suited for proteinaseous materials. The repeatability of the process and the consistency in the color content therefore recommended for Industrial application (Jothi, 2008).

Table 4C.14 Fastness properties of dyed wool yarn under conventional heating with different metal modanting with *Tagetes erecta*

Dyeing methods	Wash-perspiration-rubbing-light					
	WF^a	Per_{acidic}	Per_{basic}	Rub_{dry}	Rub_{wet}	LF^b
Wool (control)	3–4	3–4	3	3	3	3–4
Wool (alum)	4	4	4	4	4	4
Wool (FeSO$_4$)	5	4–5	4–5	4–5	4–5	5
Wool (CuSO$_4$)	4–5	4	4	4	4	4–5

[a] Wash fastness.
[b] Light fastness.

Table 4C.15 Dye exhaustion of *Tagetes* for wool under various conditions

Metal mordant	Wool (%)
Controlled	29
Alum	37
Ferrous sulfate	51
Copper sulfate	47
Stannous chloride	38
Stannic chloride	39
Pot. dichromate	48

4C.7   Wool dyeing with extract of *Nephelium Lappaceum* (Rambutan)

4C.7.1   Dyeing

It was carried out in the following manner: A two step dyeing (in the ratio of 1% or 2% mordant, owf) was used as pretreatment and then dyeing with *Rambutan* extract was carried out for 3 h at temperature of 30–40°C (Vankar et al., 2007). The wool was fixed with Sodium Chloride solution then rinsed thoroughly in tap water and allowed to dry in open air. The colorimetric data obtained from dyed wool yarn pretreated by only metal mordant reveal that pretreatment markedly improved the wash fastness, in terms of change of shade of the dyed wool with respect to control samples. It also increased the color strength and flattened the shade of the dyeing. K/S were measured for wool yarn as shown in Fig. 4C.14 and CIE lab values were shown in Table 4C.16.

It was observed that the premordanting technique with metal mordants imparted good fastness properties to the wool yarn. Control samples without mordant were also prepared for comparison. Therefore, in premordanting technique, the dyed fabrics were mordanted with stannic chloride, stannous chloride, ferrous sulfate, copper sulfate, potassium dichromate, and alum. The order of K/S values is as following: $Fe \rightarrow Cu \rightarrow K \rightarrow Sn(IV) \rightarrow Sn(II) \rightarrow Al$ in wool for *Rambutan*, the absorption of color by wool yarn was enhanced using metal mordants. Fe (II) provides best chelation in all the cases due to empty d-orbitals.

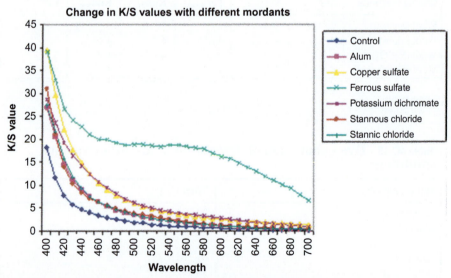

Fig. 4C.14 Change in K/S values with different mordants for wool yarn.

Table 4C.16 **L*, a*, b*, C, and H values for wool yarn dyed with** *Rambutan*

Method	Mordant	L*	a*	b*	C	H	K/S
	Control	57.567	8.23	32.88	33.89	75.91	43.26
Premordanting	Alum	57.633	8.70	33.03	34.16	75.21	79.39
	Copper sulfate	55.814	4.88	29.60	30.00	80.60	133.54
	Ferrous sulfate	46.34	2.19	5.06	5.52	6658	335.45
	Pot. dichromate	54.605	5.21	27.83	28.32	79.34	130.17
	Stannous chloride	56.154	9.91	30.16	31.74	71.78	80.30
	Stannic chloride	58.300	9.51	34.18	35.48	74.43	80.94

Table 4C.17 **Fastness properties of dyed cotton, silk fabrics, and wool yarn under conventional conditions of metal modanting with *Rambutan***

Dyeing methods	Wash-perspiration-rubbing-light					
	WF	**Per$_{acidic}$**	**Per$_{basic}$**	**Rub$_{dry}$**	**Rub$_{wet}$**	**LF**
Wool (control)	3	3	3	3	3	3
Wool (alum)	4	4	4	4	4	4
Wool (FeSO$_4$)	5	4–5	4–5	4–5	4–5	5
Wool (CuSO$_4$)	4–5	4	4	4	4	4–5

WF = wash fastness, LF = light fastness.

Thus Table 4C.17 shows that alum mordanting has caused improved fastness properties under all the three types of material as a representative case. Similar observations were made in the case of other three mordants as well. Marked improvement can be noticed in the case of washing and light fastness. Shades of brown were obtained in this dyeing as given in shade card (Fig. 4C.15).

Nephelium lappaceum (Rambutan) pericarp, which is a waste material, has been shown to have good dyeing properties. Metal mordant in conjunction with *Nephelium lappaceum* was found to enhance the dyeability and fastness properties effect in case all three types of material. Thus the net enhancement of dye uptake due to metal mordanting has been found to be ranging from 60%–65% in wool with respect to the control samples. The higher percentage of color strength in the case of wool makes Rambutan best suited for all kinds of natural materials. The two step dyeing process developed for the ease of industrial application has the potential to be accepted in industry.

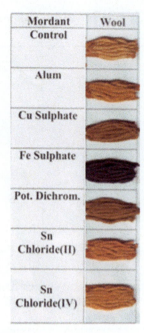

Fig. 4C.15 Shade card of Rambutan.

4C.8 Wool dyeing with *Curcuma domestica Valet* extract

4C.8.1 Dyeing

Accurately weighed textile sample was treated with 2%–4% of different metal salts such as stannic chloride, stannous chloride, alum, ferrous sulfate, copper sulfate, and potassium dichromate; only premordanting with metal salts was carried out before dyeing. Turmeric is one dye that is known to have inconsistent color changes with mordants (Vankar et al., 2008a).

The fabrics and yarn were dyed with usual method at pH 4–5 by adding acetate buffer solution at 60°C. The dyed yarn was then rinsed with water thoroughly, squeezed, and dried. The dyed material was then dipped in brine for dye fixing.

The color yield of both control and mordanted samples were evaluated by light reflectance measurements using Colorscan machine. Fig. 4C.16 shows the K/S values for curcuma dyed wool samples.

After the dyeing process was complete, the samples were dipped in brine for 1 h and then washed with plain water. This dye can be used for dyeing in various shades of yellow for wool with the use of different mordants. Alum, tin, yielded bright yellows, chrome produced yellow-gold, and iron produced shades of brown. The results are very encouraging; it is for the first time that sonicator was used for dyeing wool for using *Curcuma*. The Table 4C.18 shows L, a*, and b* values and can be seen that mordants show higher and lower value of L for lighter and deeper shades, respectively. The Fig. 4C.16 shows the

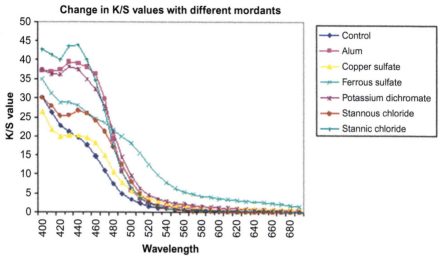

Fig. 4C.16 K/S values for curcuma dyed wool yarn.

Table 4C.18 **L*, a*, and b* values for dyed wool with Curcuma (premordanting) under conventional and sonicator conditions**

Mordant	Color obtained	L*	a*	b*	K/S	δE
Blank	White	83.500	0.90	0.82	1.73	–
Control (C)	Bright	66.655	6.54	36.06	45.30	–
Control (S)	yellow	65.659	8.83	37.66	72.44	–
Alum (C)	Yellow	67.741	8.56	37.02	138.82	2.48
Alum (S)		63.686	12.07	36.77	153.85	3.89
Stannic chloride (C)	Mustard	67.451	13.19	36.99	141.66	6.76
Stannic chloride (S)	yellow	61.845	17.99	35.52	170.62	10.14
Stannous chloride (C)	Orangish	68.299	21.52	37.93	102.86	15.18
Stannous chloride (S)	yellow	64.310	24.32	37.50	124.31	15.54
Ferrous sulfate(C)	Khaki	48.369	11.33	19.74	185.03	24.97
Ferrous sulfate (S)	green	42.664	12.05	19.15	211.96	29.69
Copper sulfate (C)	Light	55.622	6.79	26.85	109.54	14.37
Copper sulfate (S)	green	53.021	5.84	28.19	142.54	16.07
Pot. dichromate (C)	Mustard	60.235	8.87	31.01	177.93	8.49
Pot. dichromate (S)	green	53.928	11.10	29.20	185.08	14.63

differences of K/S values for unwashed and washed dyed by sonicator and conventional dyeing methods were evaluated for sonicator dyed samples of wool. The shades obtained were bright and they possessed very good fastness properties particularly the ones dyed under sonication as shown in Table 4C.18. But the main challenge for the dyers still remains attaining good fastness properties for *Curcuma* dyed fabric.

All samples dyed under sonication condition showed lower L (lightness) values as compared to conventionally dyed samples, similarly the C (chrome) values showed marked difference.

Curcuma domestica extract was used for dyeing wool yarn under conventional and sonication conditions. The summarized results of wash and light fastnesses have been shown in Table 4C.19. The dyed fabrics show improved wash and light fastnesses for sonicator dyed sample in all the cases. The K/S values also show higher values while considering the wash and light fastnesses of the dyed wool yarn. Beautiful shades of yellow were obtained by *Curcuma* dyeind as mentioned/shown in Fig. 4C.17A and B.

Table 4C.19 Wash and light fastnesses of dyed wool

Type of dyeing	Color staining based on K/S value at 450 nm			
	Conventional	Sonicator	Conventional	Sonicator
Wool	9.13	23.27	6.28	22.24

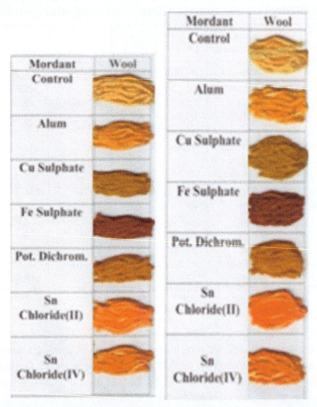

Fig. 4C.17 Shades of wool dyed by curcuma in (A) conventional method and (B) sonicator method.

Conclusion

Many Natural dyes were used for wool dyeing by various researchers like Henna (Dev et al., 2009), Lissi (Vankar and Dixit, 2011), Portulaca (Shanker et al., 2004), Mirabilis (Shanker and Vankar, 2005c), Gomphrena (Shanker and Vankar, 2005b), Gurhal (Shanker and Vankar, 2007b), and Symplocos (Vankar et al., 2008b). Wool dyeing with newer natural dye sources such as *Celosia*, *Nerium*, *Hollyhock*, *Hibiscus mutabilis*, *Caryatia*, *Tegetus*, *Rambutan*, and *Curcuma* gave very good fastness properties and dye adherence on wool. Much better colors were obtained when traditional natural dyes that have been used as these were tried and tested for hundreds of years. Some innovation at the stage of pretreatment has proven to be very good for wool dyeing; the CIE Lab values and the K/S values have been found to be well within the acceptable limits.

Furher Reading

- R. Kanchana, A. Fernandes, B. Bhat, S. Budkule, S. Dessai and R. Mohan, 2013. Dyeing of textiles with natural dyes-An eco-friendly approach. International Journal of ChemTech Research, 5(5), pp. 2102–2109.
- K.H. Prabhu and A.S. Bhute, 2012. Plant based natural dyes and mordants: A Review. J. Nat. Prod. Plant Resour, 2(6), pp. 649–664.
- D.M. Lewis, and J.A. Rippon, eds., 2013. The coloration of wool and other keratin fibres. John Wiley & Sons.
- M. Periolatto, F. Ferrero, M. Giansetti, R. Mossotti, and R. Innocenti, 2010. Enzyme-aided wool dyeing with a neutral protease at reduced temperatures. Engineering in Life Sciences, 10(5), pp. 474–479.
- S.J. McNeil and R.A. McCall, 2011. Ultrasound for wool dyeing and finishing. Ultrasonics sonochemistry, 18(1), pp. 401–406.

References

Dev, V.G., Venugopal, J., Sudha, S., Deepika, G., Ramakrishna, S., 2009. Dyeing and antimicrobial characteristics of chitosan treated wool fabrics with henna dye. Carbohydr. Polym. 75, 646–650.

Ghorpade, B., Darvekar, M., Vankar, P.S., 2000. Ecofriendly cotton dyeing with Sappan wood dye using ultrasound energy. Colourage 27.

Jothi, D., 2008. Extraction of natural dyes from African marigold flower (*Tagetes ereectal*) for textile coloration. Autex Res. J. 8, 49–53.

Popoola, A., 2000. Dyeability of cellulose fibers using dyestuff from African rosewood (*Pterocarpus erinaceous*). J. Appl. Polym. Sci. 77, 746–751.

Riva, A., Alsina, J., Prieto, R., 1999. Enzymes as auxiliary agents in wool dyeing. Color. Technol. 115, 125–129.

Sarkar, A.K., Seal, C.M., 2003. Color strength and colorfastness of flax fabrics dyed with natural colorants. Cloth. Text. Res. J. 21, 162–166.

Shanker, R., Vankar, P.S., 2005a. Dyeing with *Celosia cristata* flower on modified pretreated wool. Colourage 52, 53.

Shanker, R., Vankar, P.S., 2005b. Dyeing wool with *Gomphrena globosa* flower. Colourage 52, 35–38.

Shanker, R., Vankar, P.S., 2005c. Ultrasonic energised dyeing of wool with *Mirabilis jalpa* flowers. Colourage 52, 57–61.

Shanker, R., Vankar, P.S., 2007a. Dyeing cotton, wool and silk with *Hibiscus mutabilis* (Gulzuba). Dyes Pigments 74, 464–469.

Shanker, R., Vankar, P.S., 2007b. Dyeing wool yarn with *Hibiscus rosa sinensis* (Gurhhal). Colourage 54, 66–69.

Shanker, R., Shivani, V., Vankar, P.S., 2004. Ultrasound energised dyeing of wool with Portulaca flower extracts using metal mordants. Colourage 51, 41–46.

Tiedemann, E.J., Yang, Y., 1995. Fiber-safe extraction of red mordant dyes from hair fibers. J. Am. Inst. Conserv. 34, 195–206.

Vankar, P.S., 2009. Utilization of temple waste flower-*Tagetus erecta* for dyeing of cotton, wool and silk on industrial scale. J. Text. Appar. Technol. Manag. 6.

Vankar, P.S., Dixit, S., 2011. Natural dyeing of cotton, wool and silk with the stem and leaves extract of *Illicium griffithii*. Res. J. Text. Appar. 15, 77–83.

Vankar, P.S., Shanker, R., 2006. Dyeing cotton, silk and wool with *Cayratia carnosa* Gagn.or Vitis *trifolia*. Asian Text. J. 38.

Vankar, P.S., Shanker, R., 2008. Ultrasonic dyeing of cotton and silk with *Nerium oleander* flower. Colourage 55, 90–94.

Vankar, P.S., Shanker, R., 2006. Dyeing silk, wool and cotton with Alcea rosea flower. Fibre2Fashion.com, electronic portal.

Vankar, P., De Alwisb, A., De Silvab, N., 2007. Dyeing of dyeing cotton, wool and silk with extract of *Nephelium lappaceum* (Rambutan) pericarp. Asian Text. J. 66–70.

Vankar, P., Shanker, R., Wijayapala, S., De Alwisb, A., De Silvab, N., 2008a. Sonicator dyeing of cotton, silk and wool with *Curcuma domestica* Valet extract. Int. Dyers 193, 38–42.

Vankar, P.S., Shanker, R., Dixit, S., Mahanta, D., Tiwari, S., 2008b. Sonicator dyeing of natural polymers with *Symplocos spicata* by metal chelation. Fibers Polym. 9, 121–127.

Vankar, P.S., Shanker, R., Wijayapala, S., De Silva, N.G.H., De Alwis, A., 2009. Dyeing of cotton, wool and silk with the flower extracts of *Tegetus erecta* (Marigold). J. Text. Appar. Technol. Manag. 6.

Innovative dye extraction methods 5

D. Shukla, P.S. Vankar
FEAT (Facility for Ecological and Analytical Testing), Kanpur Kalyanpur, India

Introduction

A natural dye implies those colorants that are obtained from animal or vegetable matter without chemical processing, unlike synthetic dyes, which are synthesized from chemical predecessor. These dye-bearing materials contain only a small percentage of dye usually 0.5%–5%. These plant materials cannot be directly used for dyeing textiles.

The traditional method used to extract the dyestuffs from all other plants mentioned in the past where the plant material was added directly to the dye bath. This has been used by dyers for centuries and is still used by many dyers in northeastern states of India. The efficiency of this method is poor. Extraction of natural colorant from plants has always been a challenging issue; thus, it was found that procedures for the extraction of coloring matter in aqueous medium have to be standardized at pilot and bulk-scale plant. Tedious extraction of coloring component from the raw material, low color value, and long dyeing time push the cost of dyeing with natural dyes considerably higher than with synthetic dyes. In case of sappan wood, prolonged exposure to air converts the colorant brasiline to brazilein, causing a color change from red to brown (Vankar, 2000). The traditional method used to extract the dyestuffs from all other plants mentioned earlier, where the plant material is added directly to the dye bath. This has been used by dyers for centuries and is still used by many dyers in northeastern states of India. The disadvantages of this method are the following:

- The plant material has to be separated from the textile.
- It is not applicable to modern textile fabrication machines (pumps and spinnerets will be choked).
- Hard plant material such as madder roots or barks of *cassia* and amla is difficult to extract.
- The low density of the dried material requires high processing volume.

Disadvantage has to be solved for use by modern mills. For industrial use, the best method is to provide extracts. Aqueous extracts are not especially favorable for dye plants such as parkia, alkanet, and tulsi where 50:50 water/methanol extract for dyeing was used. The reason being flavonoids, anthraquinones, and aglycones are poorly soluble in water and therefore are get extracted only partially. The remaining material always contains a considerable amount of dyestuffs. A method to extract the dyestuffs from such plants is to boil the powdered material with methanol for 1 h. This method is used for quantitative determination of the dye content. Also, an alkaline extraction of madder gave promising results. Because of their slightly acidic character, flavonoids and anthraquinones are soluble in alkaline solutions and, after drying, also in water. This method gives good reproducible relations between the dye content and the dyeing power.

Natural Dyes for Textiles. http://dx.doi.org/10.1016/B978-0-08-101274-1.00005-7

As natural dye-bearing materials contain only a small percentage of coloring matter or dye along with a number of other plant such as water-insoluble fibers, carbohydrates, protein, chlorophyll, and tannins, among others, extraction not only is a necessary step for preparing purified natural dyes but also is required to be carried out by users of crude dye-bearing materials. As natural coloring materials are not a single-chemical entity and the plant matrix also contains a variety of nondye plant constituents, extraction of natural dyes is a complex process. The nature and solubility characteristics of the coloring materials need to be ascertained before employing an extraction process (Saxena and Raja, 2014). The different methods for extraction of coloring materials are the following:

- Aqueous extraction
- Alkali or acid extraction
- Microwave- and ultrasonic-assisted extraction
- Fermentation
- Enzymatic extraction
- Solvent extraction
- Super critical fluid extraction

To overcome drawbacks in aqueous extraction, sonicator was used and found that the dye extraction was much faster. Efficient extraction of the dye from the plant material is very important for standardization and optimization of vegetable dyes. Novelty in extraction process is straight away needed in making natural dyeing feasible at industrial level.

5.1 Innovation in extraction process

Since the last decade, the application of natural dyes on textile materials is gaining popularity all over the world, possibly because of increasing awareness of environment, ecology, and pollution control. Application of natural dyes in today's scenario makes use of modern science and technology not only to revive the traditional technique but also to improve its extraction, thereby increasing its rate of production, cost affectivity, and consistency in shades. It therefore requires some special measures to ensure evenness in dyeing.

(a) *Soxhlet extraction*—This extraction process is used for continuous solvent extraction of semivolatile analytes from solid matrices such as natural dyes. The use of an inert all-glass system for extracting semivolatile compounds from a solid or semisolid sample matrix into an organic extraction solvent, such as methanol, hexane, acetone, or methylene chloride. The Soxhlet apparatus is simple to set up and use, and it features ground glass joints for easy dismantling and cleaning. The thimble is loaded into the main chamber of the Soxhlet extractor. The extraction solvent to be used is placed in a distillation flask. The flask is placed on the heating element. The Soxhlet extractor is placed on top of the flask. A reflux condenser is placed on top of the extractor. Cold water is circulated in the distillation assembly for efficient extraction and recovery of the solvent (Fig. 5.1). The extraction of dyes from *Reseda luteola* was carried out by soxhlet apparatus. The color components extracted and isolated from weld plant were characterized by column chromatography, thin-layer chromatography (TLC), NMR, mass, and IR techniques.

Fig. 5.1 Soxhlet extraction method.

(b) *Supercritical fluid extraction*—SCFE is a two-step process, which uses a dense gas as solvent usually carbon dioxide above its critical temperature (31°C) and critical pressure (74 bar) for extraction. The natural product is powdered and charged into the extractor. Carbon dioxide is fed to the extractor through a high-pressure pump (100–350 bar). The extract charged carbon dioxide is sent to a separator (60–120 bar) via a pressure reduction valve. At reduced temperature and pressure conditions, the extract precipitates out in the separator. The extract-free carbon dioxide stream is introduced several times for effective extraction of all the dye material from the natural product (Fig. 5.2).

SCFE is superior over the traditional solvent extraction of natural dyes because it uses a clean, safe, inexpensive, nonflammable, nontoxic, environmentally friendly, and nonpolluting solvent-carbon dioxide (CO_2). Secondly, the energy costs associated with SCFE are lower than the conventional techniques. Supercritical fluid extraction of natural dye from eucalyptus bark for cotton dyeing in microwave and sonicator was carried out. SCFE is a superior technique over traditional solvent extraction for natural dyes. The eucalyptus bark is shredded off; a plant waste has been exploited for natural dyeing of cotton (Vankar et al., 2002). Bright yellow color was obtained with stannic chloride mordant that showed very good fastness both in the case of microwave and sonicator dyeing, though the latter was superior (Mirjalili et al., 2011).

(c) *Solvent extraction*—A method of liquid-liquid *extraction* is a partitioning method, to separate compounds based on their relative solubilities in two different immiscible liquids, usually water and an organic *solvent*. It is an *extraction* of a substance from one liquid into another liquid phase, or it could be from solid to liquid as in the case of natural dyes, where the colorant is extracted from solid plant part to solvent which could be either water or any

Fig. 5.2 Supercritical extraction.

other solvent. *Alkanna tinctoria* (alkanet) belongs to family Borginaceae. The root, which is often very large in proportion to the size of the plant, yields in many of the species a red dye. Alkanet root is a good source of natural color from the roots. The plant is also called *Anchusa tinctoria*. It was earlier used for soap; traditionally, the color was extracted from the root by infusing in oil, then using the oil in soap manufacture. It gave burgundy and purples in alkaline medium. So far, only nylon and polyester fibers have been dyed by alkanet. This dye has been conventionally used as a food (Tiwari and Vankar, 2001). The coloring matter in alkanet root is anchusin, a naphthoquinone derivative containing three hydroxyl groups, two of which are phenolic is primarily extracted with alcohol. The main pigment is alkannin that was earlier called anchusin. As alkanet is a naphthoquinone-based dye, theoretically, it is expected to behave as a disperse dye. The dye pigment is insoluble in water but has been used to dye wool, silk, and cotton at 40°C for 1 h with addition of alcohol. The innovative solvent extraction of the dry stem leaves of *Illicium griffithii* is an easy process; the solvent is removed, and the aqueous extract was used for dyeing. The CIELab and color strength (K/S) of the dyed fabrics were good. The superiority of solvent extraction over conventional extraction has been established through this study (Vankar and Dixit, 2011).

(d) *Hot water extraction*—Fruits of *Cayratia* plant source were crushed and dissolved in distilled water and allowed to boil in a beaker kept over water bath for quick extraction for 3 h. All the color was extracted from fruits by the end of 3 h. The extraction of anthocyanins from fruit skins is comparatively simple. The dark bluish purple skins of the berries are separated from the rest of the fruit, freed completely from pulp by hydraulic pressure, and extracted with water without delay and used for dyeing. The solution was filtered for further use. *Cayratia* shows good prospects for textile dyeing (Vankar and Shanker, 2006b).

(e) *Oil/water microemulsion extraction*—Pigment from *Canna indica* flower in different media was used; aqueous, oil/water microemulsion and ethanolic, oil/water microemulsion were prepared and used for dyeing cotton. The highest dye uptake was observed for ethanolic extract of canna flower (Vankar and Srivastava, 2010).

(f) *Citric acid/methanol extraction*—Anthocyanin from hibiscus flowers has been extracted by developing a method using methanolic solution of 4% citric acid. The new method gave better yield of anthocyanin as compared with methanolic solution of 0.1% hydrochloric acid. It has been also shown that pH of the extract plays an important role on the dye, thus by adjusting the pH of the extract (Srivastava et al., 2008).

(g) *Ultrasound or sonicator extraction*—Sonicator dyeing is very innovative technique and fuel saver methodology. In this method ultrasound energy of 20 KHz frequencies is utilized. Sonicator has high-energy sound waves, which increase ultrasonic cavitations (Fig. 5.3). This releases considerable amount of energy. This methodology is advantageous where energy resources are limited. Even heat sensitive dyes can be used in sonicator dyeing very comfortably without undergoing decomposition. The dye uptake is very good in sonicator dyeing. The same bath can be recharged and reused. The sonochemical activity arises mainly from acoustic cavitations in liquid media. The acoustic cavitations occurring near a solid surface will generate microjets; the microjet effect facilitates the liquid to move with a higher velocity resulting in increased diffusion of solute inside the pores of the fabric. In the case of sonication, localized temperature raises and swelling effects due to ultrasound may also improve the diffusion. The stable cavitations bubbles oscillate which is responsible for the enhanced molecular motion and stirring effect of ultrasound. In case of cotton dyeing, the effects produced due to stable cavitations may be realized at the interface of fabric and dye solution. Dye uptake was studied during the course of the dyeing process for a total dyeing time of 3 h with and without ultrasound. The influence of process parameters for ultrasound assisted leaching of coloring matter from plant materials. The influence of process parameters on the extraction efficiency such as ultrasonic output power, time, pulse mode, effect of solvent system, has been studied. The use of ultrasound is found to have significant improvement in the extraction efficiency of colorant. Pulse mode operation may be useful in reducing electric energy consumption in the extraction process. Ultrasound is also found to be beneficial in natural dyeing of leather with improved rate of exhaustion. Both dyed substrates have better color values for ultrasonic beet extract as inferred from reflectance measurement. Therefore, sonication clearly offers efficient extraction methodology from natural dye resources such as beetroot with ultrasound even dispensing with external heating, thereby also making eco-friendly nontoxic dyeing of fibrous substances a potential viable option (Sivakumar et al., 2009).

The color yielding plant materials used in another study included green wattle bark, marigold flowers, pomegranate rinds, 4'o clock plant flowers, and cockscomb flowers. Analytic studies such as UV/Vis spectrophotometry and gravimetric analysis were performed on the extract. The results indicate there is a significant 13%–100% improvement in the extraction efficiency of the colorant obtained from different plant materials due to the use of ultrasound. Therefore, this methodology could be employed for extracting coloring materials from plant materials in a faster and effective manner (Sivakumar et al., 2011).

For a trial to improve the natural dyeing, cultural heritage to meet the environmental future demands technology to reach high-quality dyed patterns. The work deals with extraction and dyeing of woolen fabric with *Sticta coronata* under ultrasonic energy and glucose/hydrogen peroxide-based redox system (Mansour, 2010). The efficiency of ultrasonic-assisted extraction in the presence of 9:1 water/acetone solvent and dyeing in the presence of redox system followed by alum mordanting have been studied in compared when the system was absent and the traditional thermal technique. The influence of redox system, ultrasonic energy and alum mordanting on the rate of dyeing and dye fixation and the mechanism of glucose/hydrogen peroxide redox system has been tentatively suggested. The extraction with 9:1 water/acetone solvent possesses higher absorbency in shorter extraction time compared with the aqueous one. Redox system reduced the rate of dyeing at lower temperature with significant enhancement on the dye exhaustion and fixation, involving covalent bonding in addition to the usual coulombic bond. Ultrasonic energy provided easy efficient route for dye extraction, dyeing, and mordanting processes in compared with the traditional thermal technique (Mansour, 2010). *Ixora* flowers were also extracted by sonication (Vankar and Shanker, 2006a).

Fig. 5.3 Sonicator extraction.

(h) *Alkaline extraction*—The alkaline conditions for extraction of natural dye from henna leaves were optimized, and the resulting extract was used to further optimize its dyeing conditions on cotton by exhaust method. It was found that dyeing produced with alkaline extracts of henna leaves has better color strength than the dye extracts obtained in distilled water. Furthermore, dyeing with alkaline extracts has moderate to good fastness properties and that mordanting did not result in any significant improvement in fastness properties. Finally, in comparative studies between synthetic and this natural dye, it was inferred that natural dye has good potential to act as copartner with synthetic dye (Ali et al., 2009).

(i) *Pressurized liquid extraction*—Pressurized liquid extraction (PLE) is one of the so-called green technologies. PLE combines high pressure and high temperature in order to modify the properties of the extraction solvents, allowing the selection of types of extracted metabolites according to their polarity. The small solvent volume required in PLE makes this technique highly applicable to many analytic approaches, since it can provide representative extracts of samples in an automatic way while facilitating the subsequent compound identification based on the polarity of the solvent used as extractant (Mustafa and Turner, 2011).

5.2 Innovation in extraction process for newer natural dyes

5.2.1 Preparation and optimization of aqueous extract of Salvia

The flowers of *Salvia* were found to give out color in hot water very easily. The flowers were frozen after collection and then dipped in hot boiling water to get the maximum color in 30 min, which shows deepening of hue color. Increasing the quantity of flowers from 2 to 20 g/100 mL water boiled for 60 min is accompanied with the increase in color strength and depth in color hue (Vankar and Kushwaha, 2011).

5.2.2 Extraction of the dye (Canna)

Fresh, dry, and frozen flowers were crushed and dissolved in distilled water, ethanol, O/W and at RT in a beaker kept for stirring for quick extraction for 1 h. All the color was extracted from flowers by the end of 1 h.

5.2.2.1 Optimization of aqueous extraction condition

Fresh canna flowers were completely air-dried at room temperature. The flowers were crushed and ground to make paste in distilled water and were further diluted with distilled water for dyeing purpose. Similarly, frozen flowers were used for aqueous extraction.

5.2.2.2 Extraction amount and time required

The air-dried crushed flowers (2.5, 5, 7.5, 10, and 12.5 g each) were soaked in sufficient water (app. 200–250 mL) at 30–32°C for 0.5, 0.75, 1.0, and 1.25 h. After extraction, the extract was filtered through ordinary filter paper, and the filtrate was collected, and absorbance was recorded for determination of concentration of the colorants. Similarly, frozen flowers were used for ethanolic extraction.

5.2.2.3 Preparation of micro-emulsions in Canna

The microemulsion of water/surfactant/oil (50 mL/0.2 g/0.2 g) system was prepared by adding fresh flowers (15 g) to a mixture of water, oil, and surfactant (span 80) and homogenized with a mechanical stirrer at approximately 500 rpm. The samples were kept at room temperature (30°C) to equilibrate and then filtered through whatman filter paper 42. Similarly, the microemulsion was prepared by adding water to a mixture of frozen flowers (15 g) and surfactant (span 80) and homogenized with a mechanical stirrer at approximately 500 rpm. The samples were kept at room temperature (30°C) to equilibrate (Srivastava et al., 2008).

5.2.2.4 Dye solubilization

The optimization of solubility of canna flower (15 g) was carried out in different solvents such as water (50 mL), ethanol (50 mL), and microemulsion of water/surfactant/oil (50 mL/0.2 g/0.2 g). The dye solution was filtered, and the dye dissolution was determined by measuring the visible absorbance spectrum at λ_{max} 540–542 nm. The dye bath was prepared with fresh flower aqueous extract, frozen flower aqueous extract, fresh flower ethanolic extract, frozen flower ethanolic extract, and O/W microemulsion. Dyeing of cotton fabric was carried out in sonicator bath at room temperature for 30 min.

5.2.3 Extraction of dye (Rhododendron)

Dark-red variety of rhododendron flowers from plant source were crushed and dissolved in distilled water and allowed to boil in a beaker kept over water bath for quick extraction for 3 h. All the color was extracted from flowers by the end of 3 h. The solution was filtered for further use (Vankar and Shanker, 2010).

5.2.4 Extraction of the dye (Cosmos)

Alcoholic extract of sun-dried cosmos flowers was obtained by soxhlet extraction in methanol. For aqueous extract, the frozen flowers were dipped in water for 2 h till all dye is bleached out from the flowers at room temperature or on mild heating ($\approx$50°C) (Vankar et al., 2001).

5.2.5   Extraction and purification of dye (Terminalia arjuna)

pH of extraction medium: The ground bark (10 g each) was soaked in beaker that contain water (app. 200–250 mL) of different pH (4, 7, and 9). It was filtered after cooling through ordinary filter paper, and the filtrate was collected, dried at $70\pm5°C$, and weighed to calculate percent yield of the extracted mass.

Mass-to-liquor ratio: The ground bark was soaked in four beakers having 100, 200, 300, and 400 mL at 70°C for 3 h. It was filtered through ordinary filter paper, and the filtrate was collected, dried at $70\pm5°C$, and weighed to calculate percent yield of the extracted mass.

Solubility: Water, acetone, ethanol, hexane, methanol, and ethyl acetate were used for ascertaining the solubility of extracted dye. Half a gram of the extracted dyes was taken in different conicals, and 20 mL of each solvent was added in separate conicals. The flasks were kept for stirring for 1 h at room temperature (30°C); after 1 h, solvent (soluble part of dye) was decanted and the residues (insoluble part of dye) left in the flask. It was dried at $70\pm5°C$ and weighed (Tiwari and Vankar, 2007).

5.2.6   Dye extraction of black carrot (Daucus carota)

Aqueous extraction of dye from black carrots has been carried out. A weighed amount of dye material (black carrots) was extracted with water. By the standard procedure, the ratio of mass of plant material to the volume of liquid was 1:20. First, all carrots were washed thoroughly so as to remove impurities. It was then grated and frozen in freezer for about 5 h. Aqueous extraction of the frozen grated carrots was carried out by pouring grated and frozen carrots in warm water of about 70–80°C. This process helps to release all the colorant immediately. Quantity loss due to evaporation was compensated by the addition of water at the end of the extraction period to obtain the initial volume. After 1 h, the extract was filtered through cotton and concentrated a little on water bath at temperature of 60°C. Yield of dye was determined by taking this extract of preweighed carrot. Grating and freezing helped in extraction due to the high surface and good accessibility of the plant material to be extracted, relatively short extraction time was sufficient. For the standardized extraction, an extraction period of 1 h was found to be sufficient for total extraction. Total anthocyanin amounts ranged from 50 to 15 g/kg dry matter. These findings comply with literature data of carrot fresh weights, when a dry matter content of 12%–19% (Shukla and Vankar, 2013).

5.2.7   Preparation and optimization of aqueous extract of Hibiscus mutabilis

The flowers of *Hibiscus mutabilis* were found to give out color in hot water very easily. The flowers were frozen after collection and then dipped in hot boiling water to get the maximum color in 30 min that shows deepening of hue color. Increasing the quantity of flowers from 2 to 20 g/100 mL water boiled for 60 min is accompanied with the increase in color strength and depth in color hue.

5.2.7.1   Anthocyanin extraction (Hibiscus mutabilis)

Anthocyanin extraction was carried out by two different solvents HCl and Citric acid. In the first method, anthocyanins were extracted from flowers with 0.1% HCl (v/v) in methanol for 2–3 h at room temperature, in darkness (Shanker and Vankar, 2007). The mixture was filtered on a Büchner funnel, and the remaining solids were washed with 0.1% HCl in methanol until a clear solution was obtained. The combined filtrates were dried using a rotary evaporator at 55°C. The concentrate was dissolved in DW, and the solution obtained was used for dyeing.

In the second method, anthocyanins were extracted from flowers with 4.0% citric acid (w/v) in methanol for 2–3 h at room temperature, in darkness. The mixture was filtered on a Büchner funnel, and the remaining solids were washed with 4.0% citric acid in methanol until a clear solution was obtained. The combined filtrates were dried using a rotary evaporator at 55°C. The concentrate was dissolved in DW, and the solution obtained was used for dyeing.

Anthocyanin from hibiscus was extracted using both 0.1% HCl and 4% citric acid, and it was found that citric acid gave good yield and better color too. Total anthocyanin content extracted from hibiscus flowers using different acids. The extraction of anthocyanins using ethanol acidified with citric acid (0.01%) instead of hydrochloric acid was reported (Vankar and Shukla, 2011). Ethanol would be preferred for food use to avoid the toxicity of methanolic solutions. Citric acid is less corrosive than hydrochloric acid, chelates metals, maintains a low pH, and may have a protective effect during processing (Main et al., 1978). Good yield of anthocyanin extraction from purple corncob was also reported (Yang and Zhai, 2010).

5.2.8   Dye extraction of Delonix regia

The scarlet or orange red petals up of *Delonix regia* were used as dye source (Vankar and Shanker, 2009).

Optimization of extraction condition: Flowers were completely dried at room temperature. The dry flowers were crushed and ground to make powder.

Extraction amount and time required: The dried-ground flower (25, 30, and 35 g each) was soaked in sufficient water (app. 100–150 mL) at 70–75°C for 0.5, 1.0, 1.5, and 2.0 h. After extraction, the extract was filtered through ordinary filter paper, and the filtrate was collected, and absorbance was recorded for determination of concentration.

Mass-to-liquor ratio: Thirty grams of the ground flowers were soaked in four beaker having 75, 100, 125, and 150 mL at 70°C for 1.5 h.

5.2.9   Extraction of colorant from Combretum indicum

Flowers from plant source were crushed and dissolved in distilled water and allowed to boil in a beaker kept over water bath for quick extraction for 3 h as followed in study of canna (Vankar and Srivastava, 2010). All the color was extracted from flowers by the end of 3 h. The solution was filtered for further use. The colorant showed one major peak, λ_{max} at 538 nm in the visible region for aqueous and at 520 nm for methanolic extract as shown in Fig. 5.4.

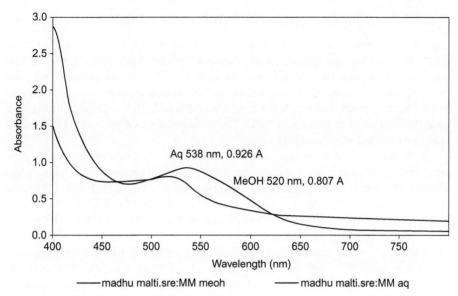

Fig. 5.4 UV/visible spectrum of *Combretum indicum*.

5.2.10 *Extraction of colorant (Nerium)*

Flowers from plant source were crushed and dissolved in distilled water and allowed to boil in a beaker kept over water bath for quick extraction for 3 h. All the color was extracted from flowers by the end of 3 h. The solution was filtered for immediate use. The flowers were also dried in trays, in thin layers, in a current of warm air immediately after picking. When dry, they are a deep, purplish black. These flowers could be used as and when required. The colorant showed one major peak, λ_{max} at 280.5 nm in the UV region (flavonoids) and at 421 nm in the visible region. The color content in fresh flowers was carried out to evaluate the color depth available for dyeing the fresh flower source. The changes in visible graph at different pH were observed. At pH 4, the λ_{max} was at 520, 570 (pH 7), and 600 nm (pH 9).

5.2.11 *Extraction of colorant in hollyhock (Alcea rosea)*

Flowers from plant source were crushed and dissolved in distilled water and allowed to boil in a beaker kept over water bath for quick extraction for 3 h. All the color was extracted from flowers by the end of 3 h. The solution was filtered for immediate use. The flowers were also dried in trays, in thin layers, in a current of warm air immediately after picking. When dry, they are a deep, purplish black. These flowers could be used as and when required. The colorant showed one major peak, λ_{max} at 296.33 nm in the UV region (flavonoids) and at 547.96 nm in the visible region. Comparison of the color content in dry and fresh flowers was also carried out to evaluate the color depth available for dyeing from the two sources. The extract shows changes in visible graph at different pH. At pH 4, the λ_{max} was at 525, 545 (pH 7), and 585 nm (pH 9) (Vankar and Shanker, 2006c).

5.2.11.1 Preparation and optimization of aqueous extract of hollyhock

The flowers of hollyhock were found to give out color in hot water very easily. The flowers were frozen after collection and then dipped in hot boiling water to get the maximum color in 30 min, which shows deepening of hue color. However, increasing the quantity of flowers from 2 to 20 g/100 mL water and boiling the extract for 60 min result in increase in color strength and depth in color hue.

5.2.12 Extraction of the dye (Tegetus)

One hundred grams of dry flowers from *Tagetes* were crushed and dispersed in ethanol (500 mL) and heated to (50°C) in a round bottom flask kept over water bath for quick extraction for 1.5 h. All the colors were extracted from flowers by the end of 1.5 h. After extraction, the extract was filtered through ordinary filter paper, the filtrate was collected, and the solvent was evaporated on rotatory evaporator and recovered to dryness. One hundred milliliters of distilled water was added to this extract. The absorbance was recorded for determination of concentration of the aqueous extract. UV-Vis spectrum of the extract showed higher color content in this case. This was further used for dyeing the fabrics.

5.2.12.1 Conventional extraction

The dried flowers of *Tagetes* (100 g) were crushed and dissolved in distilled water (500 mL) and allowed to boil in a beaker kept over water bath for extraction for 3 h. All the color was extracted from flowers of *Tagetes* by the end of 3 h. The solution was filtered, evaporated to half volume (250 mL). The extract was used for dyeing for the comparison purpose. UV-Vis spectrum of the extract showed poor color content in this case. CIELAB and K/S of the dyed fabrics were evaluated for the differently extracted dye solutions. However, both in terms of dye content and hue color, it was not satisfactory; thus, this extract was not found suitable for dyeing purpose (Vankar, 2009).

5.2.13 Extraction of the dye (Curcuma domestica)

Curcuma domestica rhizome plant source was crushed and dissolved in distilled water and allowed to boil in a beaker kept over water bath for quick extraction for 1.5 h. All the color was extracted from rhizome by the end of 1.5 h. The extraction was filtered, concentrated, and used for dyeing. The colorant showed one major peak at 465 nm; this dye is inexpensive and abundantly available, and the method of application is very simple, producing no pollutants (Vankar et al., 2008a).

5.2.14 Extraction of biomordant

Twenty grams of the dried *Pyrus pashia* fruits were ground and soaked in water 100 mL at 70°C for 2 h. The extract was filtered and used during the dyeing process as *Rubia* and *Cayratia* (Vankar et al., 2008b).

Conclusion

The concept of environmental awareness has recently had a major impact on the textile industry and on the fashion world as well. In this context, the use of natural fibers and the development of natural dyeing processes gradually became important goals of the textile industry. Usually, methods of collection and extraction of dyes are still crude and traditional with only a few experts related to industries being well versed with dyeing procedures. As such, plenty of materials are improperly exhausted in the procedure. Therefore, proper utilization requires understanding of sustainability as well as specific preference of use pattern. Indigenous traditional and current innovative knowledge on various resources including dye-yielding plants and their optimum extraction is very essential for overall development of natural dyeing industry.

Further Reading

- Kamel M, El shishtawy R, Yousef BM, Mashaly HM (2005) Ultrasonic assisted dyeing III. "Dyeing of wool with lac as a natural dye" Dyes Pigments 63:103–110.
- Padma Vankar S and R. Shanker, (2005) Ultrasonic energized dyeing of wool with *Mirabilis jalapa* flowers. Colourage 52(2):57–61.
- Bendak A (1989) Low-temperature dyeing of protein and polyamide fibres using a redox system. Dyes Pigments 11(3):233–242.
- Uddin, M. G. 2015. Extraction of eco-friendly natural dyes from mango leaves and their application on silk fabric. Textiles and Clothing Sustainability, 1, 7.
- Redfern, J., Kinninmonth, M., Burdass, D., & Verran, J. (2014) Using Soxhlet Ethanol Extraction to produce and Test Plant Material (Essential Oils) for their Antimicrobial Properties. Journal of Microbiology & Biology Education, 15(1), 45–46.

References

Ali, S., Hussain, T., Nawaz, R., 2009. Optimization of alkaline extraction of natural dye from Henna leaves and its dyeing on cotton by exhaust method. J. Clean. Prod. 17, 61–66.

Ghorpade, B., Darvekar, M., Vankar, P.S., 2000. Ecofriendly cotton dyeing with Sappan wood dye using ultrasound energy. Colourage 47, 27–30.

Main, J., Clydesdale, F., Francis, F., 1978. Spray drying anthocyanin concentrates for use as food colorants. J. Food Sci. 43, 1693–1694.

Mansour, H.F., 2010. Environment and energy efficient dyeing of woollen fabric with sticta coronata. Clean Techn. Environ. Policy 12, 571–578.

Mirjalili, M., Nazarpoor, K., Karimi, L., 2011. Eco-friendly dyeing of wool using natural dye from weld as co-partner with synthetic dye. J. Clean. Prod. 19, 1045–1051.

Mustafa, A., Turner, C., 2011. Pressurized liquid extraction as a green approach in food and herbal plants extraction: a review. Anal. Chim. Acta 703, 8–18.

Saxena, S., Raja, A., 2014. Natural dyes: sources, chemistry, application and sustainability issues. In: Roadmap to Sustainable Textiles and Clothing. Springer International, Hong Kong.

Shanker, R., Vankar, P.S., 2007. Dyeing cotton, wool and silk with *Hibiscus mutabilis* (Gulzuba). Dyes Pigments 74, 464–469.

Shukla, D., Vankar, P.S., 2013. Natural dyeing with black carrot: new source for newer shades on silk. J. Nat. Fibers 10, 207–218.

Sivakumar, V., Anna, J.L., Vijayeeswarri, J., Swaminathan, G., 2009. Ultrasound assisted enhancement in natural dye extraction from beetroot for industrial applications and natural dyeing of leather. Ultrason. Sonochem. 16, 782–789.

Sivakumar, V., Vijaeeswarri, J., Anna, J.L., 2011. Effective natural dye extraction from different plant materials using ultrasound. Ind. Crop. Prod. 33, 116–122.

Srivastava, J., Seth, R., Shanker, R., Vankar, P.S., 2008. Solubilisation of red pigments from Canna Indica flower in different media and cotton fabric dyeing. Int. Dyer 193(1), 31–36.

Tiwari, V., Vankar, P., 2001. Unconventional natural dyeing using microwave and sonicator with alkanet root bark. Colourage 48, 25–28.

Tiwari, V., Vankar, P.S., 2007. Standardization, optimization and dyeing cotton with *Terminalia arjuna*. Int. Dyer 6, 31–33. 35–36.

Vankar, P.S., 2000. Chemistry of natural dyes. Resonance 5, 73–80.

Vankar, P.S., 2009. Utilization of temple waste flower-*Tagetes erecta* for dyeing of cotton, wool and silk on industrial scale. J. Text. App. Technol. Manag. 6, 1–15.

Vankar, P.S., Dixit, S., 2011. Natural dyeing of cotton, wool and silk with the stem and leaves extract of *Illicium griffithii*. Res. J. Text. Appar. 15, 77–83.

Vankar, P.S., Kushwaha, A., 2011. Salvia splendens: a source of natural dye for cotton and silk fabric dyeing. Asian Dyers, Dec–Jan(6), 29–32.

Vankar, P., Shanker, R., 2006a. Sonicator dyeing of cotton and silk fabric by *Ixora coccinea*. Asian Text. J. 2, 77–80.

Vankar, P.S., Shanker, R., 2006b. Dyeing cotton, silk and wool with *Cayratia carnosa* Gagn. or *Vitis trifolia*. Asian Text. J. 38, 38–45.

Vankar, P.S., Shanker, R., 2006c. Dyeing silk, wool and cotton with *Alcea rosea* flower. Fibre 2 Fashion.

Vankar, P.S., Shanker, R., 2009. Eco-friendly pretreatment of silk fabric for dyeing with *Delonix regia* extract. Color. Technol. 125, 155–160.

Vankar, P.S., Shanker, R., 2010. Natural dyeing of silk and cotton by Rhododendron flower extract. Int. Dyers 7, 37–40.

Vankar, P.S., Shukla, D., 2011. Natural dyeing with anthocyanins from *Hibiscus rosa sinensis* flowers. J. Appl. Polym. Sci. 122, 3361–3368.

Vankar, P.S., Srivastava, J., 2010. Ultrasound-assisted extraction in different solvents for phytochemical study of *Canna indica*. Int. J. Food Eng. 6, 1–11.

Vankar, P.S., Tiwari, V., Ghorpade, B., 2001. Microwave dyeing of cotton fabric-*Cosmos sulphureus* and comparison with sonicator dyeing. Can. Text. J. 31, 31–33.

Vankar, P.S., Tiwari, V., Ghorpade, B., 2002. Supercritical fluid extraction of natural dye from eucalyptus bark for cotton dyeing in microwave and sonicator. Can. Text. J. 32, 31–35.

Vankar, P., Shanker, R., Wijayapala, S., De Alwisb, A., De Silva, N., 2008a. Sonicator dyeing of cotton, silk and wool with *Curcuma domestica* Valet extract. Int. Dyers 193, 38–42.

Vankar, P.S., Shanker, R., Mahanta, D., Tiwari, S., 2008b. Ecofriendly sonicator dyeing of cotton with *Rubia cordifolia* Linn. using biomordant. Dyes Pigments 76, 207–212.

Yang, Z., Zhai, W., 2010. Optimization of microwave-assisted extraction of anthocyanins from purple corn (*Zea mays* L.) cob and identification with HPLC–MS. Innovative Food Sci. Emerg. Technol. 11, 470–476.

Unique pretreatments and posttreatments for natural dyeing

D. Shukla, P.S. Vankar
FEAT (Facility for Ecological and Analytical Testing), Kanpur Kalyanpur, India

Introduction

Dyes can generally be described as colored substances that have affinity to the substrates to which they are being applied (Pereira and Alves, 2012). Ever since prehistoric time, man has been fascinated to color the objects of daily use employing inorganic salts or natural pigments of vegetable, animal, and mineral origins. These coloring substances, known as dyes, are the chemical compounds used for coloring fabrics, leather, plastic, paper, food items, cosmetics, etc. and to produce inks and artistic colors. Dyes are of two types, i.e., synthetic and natural. Synthetic dyes are based on petroleum compound, whereas natural dyes are obtained from plant, animal, and mineral matters (Singh and Bharati, 2014). Most of the natural dyes found in the plant kingdom, available from a number of root, berries, bark, leaves, and wood. Natural dyes have color due to the presence of coloring entities, which needs to be fixed to fabric on which dyeing is carried out. Dyeing or coloring in itself can be a complex subject to deal with as there are so many components that can affect desired end product/color on fabric or material. Many factors are responsible for dyeing of fabric such as fabric/material type, compatibility of dye and material, pretreatment of material, pH of dye bath, and mordanting.

6.1 Pre and post treatments

Natural dyeing requires certain types of pretreatments and post treatments. Pretreatment of fabric prior to dyeing mainly involves a combined process generally consisting of preparation of fabric/material for dyeing and to increase dyeability, which leads to favorable properties, such as washfastness. Pretreatment of cotton involves desizing, scouring, and bleaching. Firstly, the fabric has to be desized. Desizing involves impregnation of the fabric with the desizing agent, allowing the desizing agent to degrade or solubilize the size material and finally to wash out the degradation products. They are selected on the basis of type of fabric, eco-friendliness, ease of removal, cost of desizing agents, and its effluent management. Thus, desizing helps in removing the starch material from the fabric, in increasing the absorbency power of the fabric, in increasing the affinity of the fabric to the dry chemicals, in making the fabric suitable for the next process, in increasing the luster of the fabric, and finally in increasing of dyeing and printing ability of the fabric.

Natural Dyes for Textiles. http://dx.doi.org/10.1016/B978-0-08-101274-1.00006-9

The major desizing processes practiced are the following:

Enzymatic desizing of starches on cotton fabrics; oxidative desizing using sodium or potassium persulfates; acid desizing using dilute sulfuric or hydrochloric acids and removal of water-soluble sizes by using wetting agents and mild alkalis (http://www.handprintingguiderajasthan.in/).

6.1.1   Preparation of the cotton fabric

Cotton is natural fibers obtained from plants. Cotton is a polysaccharide composed of glucose units attached to one other in a very rigid structure. The presence of three polar hydroxyl (–OH) groups per glucose repeating unit provides multiple sites for hydrogen bonding to ionic and polar groups in dye molecules. Cotton has to undergo pretreatment to get beautiful shades from natural dyes. The fabric was desized in liquor containing 5 g of soap and 0.1% HCl/L of water. The material to liquor ratio was taken as 1:40. The fabric was boiled at 95°C for 1 h and rinsed thrice in cold water and dried under shade. The desized cotton fabric was treated with tannic acid solution. The material to liquor ratio was 4% (owf). The fabric was soaked in the tannic acid solution for 4–5 h and then air-dried.

6.1.2   Preparation of silk and wool

Silk and wool are protein obtained from insect and animals. Silk is protein fiber consisting of viscous fluid excreted from the glands of silk worm.

6.1.2.1   Degumming of silk

The process of removal of natural gum from silk is known as degumming. Since presence of gum acts as a protective layer, it also acts as natural sizing agent. Therefore, degumming of silk is carried out in fabric form after weaving. However, if the silk yarn is to be dyed, then degumming of silk in yarn form is carried out before yarn dyeing. Degumming silk is usually carried out under mild alkaline conditions using soap. However, recently protease enzymes are also suggested for silk degumming.

Wool is a naturally occurring polymer made up of amino acid repeating units. Many of the amino acid units have acidic or basic side chains that are ionized (charged). The presence of many charged groups in the structure of wool provides excellent binding sites for dye molecules, most of which are also charged. The scouring of silk fabric and wool yarn were washed with solution containing 0.5 g/L sodium carbonate and 2 g/L nonionic detergent (labolene) solution at 40–45°C for 30 min, keeping the material to liquor ratio at 1:50. The scoured material was thoroughly washed with tap water and dried at room temperature. The scoured material was soaked in clean water for 30 min prior to dyeing or mordanting.

6.1.3   Treatment of fabric before dyeing

After removing the impurity of cotton then, it is treated with 4% (owf) solution of tannic acid in water. The fabric should be dipped in tannic acid solution for at least 4–5 h.

It is squeezed and dried. Premordanting was used for fabric that is already treated with tannic acid, mordanted (2% for alum and 1% for stannous chloride, stannic chloride, ferrous sulfate, copper sulfate, and potassium dichromate separately) solution and is kept on water bath at 50°C for 1 h. It is squeezed and dried.

6.1.3.1   Importance of pretreatment processes

The causes of preparatory or pretreatment processes are as follows:

- To remove natural and added impurities.
- To impart certain desirable properties (water absorbency).
- To improve the appearance of fabric (whiteness).
- To make it suitable for subsequent processes like dyeing and printing finishing.
- Removal impurities to the maximum extent with minimum effect on fabric strength. In the case of cotton, following chemical reactions are involved while removing the impurities.
- Hydrolysis.
- Oxidation.

6.1.4   Microwave dyeing techniques for cosmos and other dyes

Microwave dyeing takes into account only the dielectric and the thermal properties. The dielectric property refers to the intrinsic electrical properties that effect dyeing by dipolar rotation of the dye and the influence of microwave field upon dipoles. The aqueous solution of the dye extract has two components, which are polar. In the high-frequency microwave field, oscillating at 2450 MHz, it influences the vibrational energy in the water molecules and the dye molecules. This causes frictional heating, while materials other than water may be dipolar or may behave as dipoles due to the stress of the dielectric field, water usually dominates, probably because it is pervasive and at high concentrations of dye mixtures (Vankar et al., 2001a,b; Tiwari and Vankar, 2001). The choice of microwave is based on this fact that attainment of desired temperature is faster in microwave as compared with conventional method.

The heating mechanism is through ionic conduction, which is a type of resistance heating. Depending on the acceleration of the ions through the dye solution, it results in collision of dye molecules with the molecules of the fiber. The mordants help and affect upon the penetration of the dye and also on the depth to which the penetration takes place in the fabric. This makes microwave superior to conventional dyeing techniques.

Microwave dyeing gives an even dye finish with a lustrous look and an even pattern of color. The advantages of dyeing in the microwave are the following:

1. It is environmentally sound as one uses much less liquid and thus can exhaust dyes or save them and have no liquid dye waste to get rid of.
2. It consumes less power.
3. Dyeing is quick as it takes several minutes only.
4. Redyeing to a desired shade is possible easily.

As compared with conventional dyeing techniques that require long hours of heating/ boiling, microwave dyeing of cotton requires only 5–10 min.

6.1.5   Sonicator dyeing technique

Similarly, in sonicator (Tiwari et al., 2000b,c; Ghorpade et al., 2000b; Shanker and Vankar, 2006), there is an activated state that causes the chemical reaction between the fabric and the dye faster. Dye uptake (exhaustion) shows (Fig. 6.1) better dye uptake by sonicator method as compared with microwave, although the latter is faster. The high rate of dyeing and high dye uptake of cotton by sonicator could be attributed to

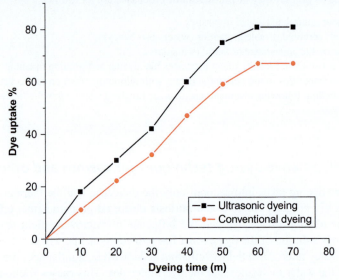

Fig 6.1 Comparison of ultrasonic dyeing with conventional dyeing in hollyhock.

the less compact structure of cotton as well. The ease of penetration of the dye into the fabric and its affinity for the dye is also dependent on metal chelation.

6.1.5.1   Effect of ultrasound

Generally, the sonochemical activity arises mainly from acoustic cavitation in liquid media. The acoustic cavitation occurring near a solid surface will generate microjets, the microjet effect facilitate the liquid to move with a higher velocity resulting in increased diffusion of solute inside the pores of the fabric. In the case of sonication, localized temperature raise and swelling effects due to ultrasound may also improve the diffusion.

It is well documented (Ghorpade et al., 2000a); the ultrasound energy gives rise to acoustic cavitations in liquid media. The acoustic cavitations occurring near a solid surface that generates microjets that has this microjet effect facilitate the liquid to get agitated with a high speed resulting in increased diffusion of solute inside the pores of the fabric. This causes localized rise in temperature and pressure, thus swelling in fiber takes place, this may also contribute in improved diffusion. The stable cavitations bubbles oscillate that is responsible for the enhanced molecular motion and stirring effect of ultrasound. In case of cotton dyeing, the effects produced due to stable cavitations may be realized at the interface of leather and dye solution. Dye uptake in the case of

Hollyhock was 81% and 67%, respectively, during the course of the dyeing process for a total dyeing time of 1 h with and without ultrasound.

6.1.6 Pre treatments in dyeing *(Rubia cordifolia)*

Dyeing was carried out in two ways:

(i) Two-step dyeing (in the ratio of 2% biomordant, owf) was used as pretreatment, and then, dyeing with *Rubia* extract (10%, owf) was carried out for 3 h at temperature 30–40°C. The dyed fabrics were rinsed thoroughly in tap water and allowed to dry in open air (Vankar et al., 2008b) (Fig. 6.2).

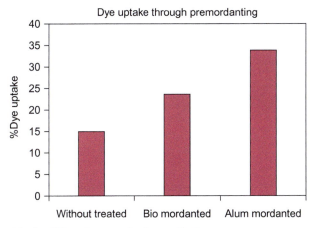

Fig. 6.2 Dye uptake by different premordanting methods.

(ii) One-step dyeing (in the ratio of 10% *Rubia* extract and 2% biomordants) were mixed thoroughly in one bath, and the moist fabric was dipped for 3 h at temperature 30–40°C. The dyed fabrics were rinsed thoroughly in tap water and allowed to dry in open air (Fig. 6.3).

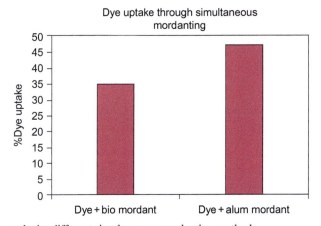

Fig. 6.3 Dye uptake by different simultaneous mordanting methods.

In case of cotton dyeing, the effects produced due to stable cavitation may be realized at the interface of cotton fabric and dye solution. Dye uptake was studied during the course of the dyeing process for a total dyeing time of 3 h with and without ultrasound. About 58% exhaustion of dye (*Rubia*) can be achieved in 3 h dyeing time using ultrasound, while, in the absence of ultrasound in stationary condition for this natural dye, only 40% was observed as shown in Fig. 6.4.

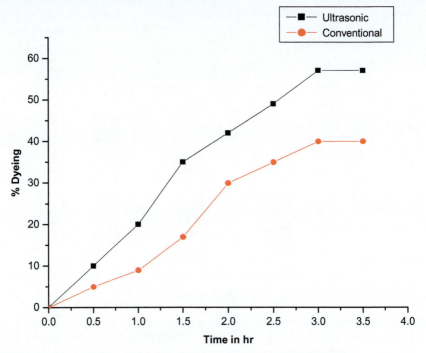

Fig. 6.4 Ultrasonic dyeing with Rubia with *Eurya acuminate* as biomordant.

6.1.7   Innovation in dyeing: value addition on dyeing

(a) *Antimicrobial dyeing*. Optimization of dyeing conditions of unripe *Citrus unshiu* extract on silk fabric and antimicrobial activity of the dyed fabric for its potential use as a functional natural dye was carried out. Unripe fruits of *Citrus unshiu* in Jeju Island, Korea, extracted in 80% ethanol solution to final solid dye powder were dyed on silk fabric under a variety of conditions such as dye bath concentrations, temperature, and dyeing duration together with mordanting. Dyeing fastness properties to washing, rubbing, perspiration, and light were tested, and the antimicrobial activities of the dyed fabric against *Staphylococcus aureus* and *Klebsiella pneumoniae* were investigated quantitatively. From these results, it was concluded that the dye of unripe *Citrus unshiu* could be employed to produce antimicrobial silk fabric. Furthermore, to get more saturated shades on the fabric by the citrus (Lee et al., 2010).

The extract of the *Berberis vulgaris* can be considered as a natural dye of acceptable fastness properties together with excellent antibacterial activity for woolen textiles (Haji, 2010).

Antimicrobial activity of wool fabric dyed with this dye was tested according to diffusion agent. Test organisms as *Escherichia coli*, *Bacillus subitilis*, *Pseudomonas aeruginosa*, and *Staphylococcus aureus* were used, and the results indicated that the samples exhibited a high inhibition zone. According to the available literature, this is the first report concerning a natural dye for fabric from fruits of red prickly pear plants (Ali et al., 2011).

(b) *New source for hair dyeing.* Indigo carmine has been used as a source of blue dye for wool and hair dyeing. The option to use indigo carmine in combination with other natural dyes in a one-bath procedure as a hybrid dyeing concept is of interest both for natural dyeing and for coloration of hair. The study showed that indigo carmine dyeing on wool exhibited substantial sorption in the range of pH 4–5 and temperature between 40 and 60°C. Experiments with human hair samples indicated that the experiments on wool could serve as a model that can be applied for hair dyeing. Comparisons of the energy, water, and chemical consumption between two-bath dyeing with indigo and natural dyes separately and a one-bath dyeing using indigo carmine together with other natural dyes demonstrated the advantages of the one-bath hybrid dyeing concept (Komboonchoo and Bechtold, 2009).

(c) *Enzymatic and ultrasonic dyeing.* Natural pigments from *Crocus sativus* stigmas were used for the dyeing of cotton and wool fibers after treatment with the enzymes α-amylase and trypsin, respectively. The separation of various compounds and pigment constituents of the stigmas was effected by fractionation of the methanolic extract on a silica gel column, and the use of these fractions for dyeing is described. The dyeing properties of the fractions were compared with those of commercial trans-β-carotene. Wash- and lightfastness of the dyed samples after the enzymatic pretreatment were also studied (Liakopoulou-Kyriakides et al., 1998).

Cotton and wool fabrics were dyed with the natural dyes chlorophyll and carmine after treatment with the enzymes cellulase, α-amylase, and trypsin. Wash- and lightfastnesses of the dyed samples were studied. Enzymatic pretreatment resulted in an increase in pigment uptake in all cases compared with the corresponding untreated samples and did not affect fastness properties. Pretreatment with metallic salts and dyeing of pretreated samples was also carried out, and the fastness properties of the dyed materials were studied. The effect of conventional mordanting with metallic salts was compared with that of enzymatic treatment on the dyeing properties of the dyes used (Tsatsaroni and Liakopoulou-Kyriakides, 1995).

Two step ultrasonic dyeing of cotton and silk fabrics with natural dyes, *Terminalia arjuna*, *Punica granatum*, and *Rheum emodi* have been developed in which an enzyme is complexed with tannic acid first as a pretreatment. This was found to be comparable with one step simultaneous dyeing. The effectiveness of three enzymes—protease-amylase, diasterase, and lipase was determined. The enzymatic treatment gave cotton and silk fabrics rapid dye adsorption kinetics and total higher adsorption than untreated samples for all the three dyes. The CIE Lab values also showed improvement by enzymatic treatment. The tannic acid-enzyme-dye combination method offers an environmentally benign alternative, "soft chemistry" to the metal mordanted natural dyeing (Vankar et al., 2007c).

The study of the natural pigment from sappan wood was used for the dyeing of wool fabrics after treatment with the protease and transglutaminase. The influences of protease and transglutaminase on the UV/visible absorption spectrum of aqueous extract of sappan were studied. The enzymatic modified wool was compared with nonmodified wool in K/S value and fastness after direct dyeing and mordant dyeing. It was shown that protease and transglutaminase made the absorbance at the λ_{max} 540 nm in visible region increase. It suggested that there might be some interaction between the enzymes and sappan dye and the residual enzyme on wool fabric might affect the color of following dyeing. Compared with untreated wool, treatments with protease and transglutaminase enhance K/S value of wool dyed subsequently with sappan. Modification of protease led to some decrease in wet rubbing fastness, whereas transglutaminase had almost no influence on rubbing fastness. Enzymatic treatments have no influence on the washing fastness for samples dyed with sappan (Zhang and Cai, 2011).

In recent years, the use of low-environmental-impact biotechnology giving rises to new types of treatment in the textile industry. The use of protease enzymes to improve some physical and mechanical properties such as smoothness, drape ability, dyeing affinity, and water absorbency is particularly interesting. In this research, wool yarns were first treated with different concentrations of protease enzymes in water solution including 1%, 2%, 4%, and 6% owf for 60 min. The dyeing process was then carried out on the treated yarns with madder (50% owf). Tensile strength of treated yarns was decreased due to enzyme treatment, and it continued to decrease with an increase in enzyme concentration in solution. The L values decreased for the samples treated with enzyme. The wash- and lightfastness properties of samples were measured according to ISO 105-CO5 and daylight ISO 105-BO1. The washing fastness properties of treated samples were not changed. In the case of lightfastness properties, it was increased a little for 4% and 6% enzyme-treated samples (Parvinzadeh, 2007).

The environmental safe treatment methods were used to improve the penetrability of wool fibers with natural dye (Madder) in the presence of chem and phys modified natural soybean lecithin. The dyeing conditions of wool fiber with madder were optimized by response surface methodology (RSM). The five independent variables were selected at low- and high-level values. The ANOVA results of the quadratic model show that the model terms such as dyeing temp, dyeing time, lecithin concentration, plasma treatment duration, dyeing method, and their interactions are significant. Moreover, the optimum conditions that were proposed for the natural dyeing of LTP-treated wool cause dyeing temperature reach to 76.53°C in the presence of acetylated lecithin (Barani and Maleki, 2011).

Polyacrylonitrile fiber cured by enzyme was dyed by natural dye cochineal. The effect of pretreating agent and enzyme concentration, temperature, time, and pH value was discussed. The conclusion was that the dye uptake of polyacrylonitrile fiber pretreated by benzyl alcohol was higher of that pretreated by ethylene glycol methyl ether, the dye uptake of polyacrylonitrile fiber cured by enzyme was highest when enzyme concentration was 5%, and enzyme cure time was 50 h,, and enzyme cure temperature was 40°C, and pH value was 7. The water return rate of polyacrylonitrile fiber cured by enzyme was increased note worthily (Gong and Ma, 2011).

A single-bath dyeing of linen fabric with natural dyes, *Curcuma longa*, logwood, and pomegranate has been developed. In this process, one or more enzymes are complexed with tannic acid and a natural dye. On the basis of the developed enzyme complexes of the natural dyes, a comparison was made between the three dyeing methods: exhaustion, pad-dry, and pad-dry-cure methods. The effectiveness of the four enzymes, neutral cellulase, protease, α-amylase, and lipase, and three mixtures of these enzymes, α-amylase + lipase, neutral cellulase + lipase, and neutral cellulase + α-amylase, was determined. A broad variation in shade, hue, and color depth can be achieved by applying enzymes and mixtures of enzymes complexes in the three dyeing methods. Colorimetric data show dyeing improvement using these enzymatic complexes. It was found that each of the four enzymes and their mixtures were very effective, when used in conjunction with tannic acid in improving the washing fastness of the three natural dyes. The use of enzyme/tannic acid/natural dye complexes replaces metal mordants making natural dyeing process more eco-friendly (El-Zawahry, 2009).

Eco-friendly ultrasonic textile dyeing with natural dyes such as Acacia catechu and *Tectona grandis* show better and faster dye uptake after enzyme pretreatment on cotton fabric, and results of dyeing are better than metal mordanted fabric. It is observed that there is marked improvement in washfastness and lightfastness. The role of enzyme pretreatment is primarily for better absorbency, adherence, and dye ability of these dyes on cotton fabric, thereby completely replacing metal mordants with enzyme for adherence of natural dyes on cotton. Scanning electron microscopy showed surface characteristics at different stages of dyeing. The effect of sonication on the dyeing is compared with conventional heating. The study also shows enhancement in CIE Lab values (Vankar and Shanker, 2008).

The dyeing properties of cotton and hosiery material with Al root by using sonicator dyeing have been reported. Aqueous extract of Al root gave very deep reddish brown color with tin mordant (Tiwari et al., 2001).

The extractability of lac dye from natural origin using power ultrasonic was also evaluated in comparison with conventional heating. The results of dye extraction indicate that power ultrasonic is rather effective than conventional heating at low temperature and short time. The effects of dye bath pH, salt concentration, ultrasonic power, dyeing time, and temperature were studied, and the resulting shades obtained by dyeing with ultrasonic and conventional techniques were compared. Color strength values obtained were found to be higher with ultrasonic than with conventional heating. The results of fastness properties of the dyed fabrics were fair to good. Dyeing kinetics of wool fiber with lac dye using conventional and ultrasonic conditions was compared. The time/dye-uptake isotherms are revealing the enhanced dye uptake in the second phase of dyeing (diffusion phase). The values of dyeing rate constant, half time of dyeing and standard affinity, and ultrasonic efficiency have been calculated and discussed (Kamel et al., 2005).

The dyeing of cationized cotton fabrics with lac natural dye has been studied using both conventional and ultrasonic techniques. The effects of dye bath pH, salt concentration, ultrasonic power, dyeing time, and temperature were studied, and the resulting shades obtained by dyeing with ultrasonic and conventional techniques were compared. Color strength values obtained were found to be higher with ultrasonic than with conventional heating. The results of fastness properties of the dyed fabrics were fair to good. Dyeing kinetics of cationized cotton fiber with lac dye using conventional and ultrasonic conditions were compared (Kamel et al., 2007).

An ultrasound-energized dyeing technique of terricot and cotton fabric with lac dye has been developed. Pretreatment of the fabrics with mild acids followed by dyeing and mordanting gave very good fastness properties to the fabric. Terricot fabric dyeing for maximum dye uptake and complete dye utilization was also studied (Tiwari et al., 2000a).

Evaluation the efficiency of ultrasonication on new natural dye obtained from leaves and stem extracts of *Daphne papyraceae* using metal mordant for good cotton, silk, and wool dyeing was carried out. It also proposes to effect the characterization of the colorant. Design/methodology/approach—for effective natural dyeing with leaves and stem extracts of *D. papyraceae*, both conventional and sonication methods for cotton, silk, and wool dyeing were carried out using (Vankar et al., 2009).

The efficiency of ultrasonication on new natural dye obtained from leaves extract of *Acer pectinatum wallich* using metal mordant for good cotton dyeing prospects was conducted. For effective natural dyeing with leaves extract of *Acer pectinatum wallich*, both conventional and sonication methods of dyeing were carried out using metal mordants. The purpose of using sonication was for improvement of dye uptake, improved dye adherence, and good wash- and lightfastnesses. Results show marked improvement by the chosen dyeing method (Vankar et al., 2008a).

Bischofia javanica Bl. (local name maub) belonging to family Euphorbiaceae for natural dye production for textile dyeing. In the study, innovative sonicator dyeing with *Bischofia* has been shown to give good dyeing results. Pretreatment with 1%–2% metal mordant and using 5% of plant extract (owf) is found to be optimum and shows very good fastness properties for cotton, wool and silk-dyed fabrics. The net enhancement of dye uptake due to metal mordanting and sonication was found to range from 30% to 50% (cotton), 37% to 52% (silk), and 38% to 50% (wool) (Vankar et al., 2007a).

The dyeing of cotton fabric using *Eclipta alba* as natural dye has been studied in both conventional and sonicator methods. The effects of dyeing show higher color strength values obtained by the latter. Dyeing kinetics of cotton fabrics were compared with both methods. The time/dye uptake reveals the enhanced dye uptake showing sonicator efficiency. The results of fastness properties of the dyed fabrics were fair to good. CIE Lab values have also been evaluated (Vankar et al., 2007b).

The dyeing of cationized cotton fabric with 3-chloro-2-hydroxypropyltrimethylammonium chloride 69% (Quat 188) using cochineal dye was studied using both conventional and ultrasonic techniques. Factors affecting dye extraction and dye bath exhaustion were investigated. The results indicated that the dye extraction by ultrasound at 300W was more effective at lower temperatures and times than conventional extraction. Also, the color strength values obtained were found to be higher with ultrasound than with conventional techniques. However, the results showed that the fastness properties of the dyed fabrics with ultrasound are similar to those of the conventional dyed fabrics. The scanning electron microscope (SEM) and x-ray diffraction (XRD) were measured for cationized cotton fabrics dyed with both conventional and ultrasound techniques, thus showing the sonicator efficiency (Kamel et al., 2011).

(d) *Ultra violet protection through dyeing.* Natural dyes are eco-friendly, harmless, and nontoxic in nature. Some natural dyes provide ultraviolet protection to the wearer. *Garcinia indica* called as an "Indian Spice and Kokum" plant. From *Garcinia indica* waste bark, dye components were extracted in neutral media. Optimization of conditions for extraction of dye and its effect on various parameters like extraction time, dye material concentration, pH, temperature, dyeing time, and temperature and dye concentration had been thoroughly investigated. Extraction and dyeing of organic cotton fabric had been carried out by using *Garcinia indica* bark extract by simultaneous mordanting techniques using alum, pomegranate rind, myrobalan fruit rind, and used tea leaves powder as a mordant. Ultraviolet protection factor testing shows that *Garcinia indica* extract dye has got excellent ultraviolet protection (Hegde et al., 2011).

(e) *pH-mediated dyeing.* A study was carried out to optimize the dyeing conditions for silk fabrics using a new natural dye, vervain barks at different pH values in the absence of mordants. Also, aspects are implemented to investigate the effects of processing conditions on color difference, color scale, color strength, and lightfastness characteristics. The changes in the morphology of the silk fabric surface after dyeing are studied using a scanning electron microscope. The pH in the acidic range of 1–5 is optimum for dyeing, showing very good fastness properties for silk-dyed fabrics (Ali and El-Mohamedy, 2011).

The effect of pH value on color of purple sweet potato solution was studied. The dyeing behavior of natural dyestuff, extracted from purple sweet potato, on wool fabric was carried out. The effects of mordants and dyeing methods were investigated, and the air permeability, breaking strength, and anti-UV properties of the dyed fabric were tested. Results showed that at pH ≤ 3 the color of purple sweet potato natural dyes solutions was red and, with the increase of pH value, the color of natural dyestuff solution changed from red to blue. The good effect was achieved after pad dyeing using rare earth as the mordant, the air permeability, and the breaking strength of the dyed fabric dropped in a certain degree, but the anti-UV property improved markedly (Li and Zhao, 2010).

(f) *Alkaline dyeing.* *Carthamus tinctorius* belongs to family Asteraceae; it yields yellow-red dye. The pigment carthamin is a mixture of quinchalcone and precarthamin. Dyeing of cotton, silk, and wool showed excellent fastness properties. Dye was extracted from the petals of the flower by alkaline solution maintaining the pH between 8 and 9. Thus, safflower dye has potentials for commercial natural dyeing and can be considered as safe alternative for azoic dye direct yellow (Vankar et al., 2004).

(g) *Morpholine pretreated dyeing*. Deep-red *Celosia cristata* was used for the dyeing of diamine-pretreated wool showed darker shade as compared with morpholine and sodium hydroxide yellow to green (Shanker and Vankar, 2005).

6.1.8   Innovation in dye fixing

Selective premordanting (single and double) and natural dyeing of 6% H_2O_2 (50%) bleached jute fabric have been carried out using myrobolan (harda) and metallic salts (potash alum and aluminum sulfate) as mordants and aqueous extract of tesu (palash flower petals) as dyeing agent under varying dyeing condition to optimize the dyeing process variables. It is found that the 20% myrobolan followed by 20% aluminum sulfate in sequence is a most potential double premordanting system rather than using them as single mordant separately, considering the results of important textile-related properties and color yield. Effects of dyeing process variables (time, temperature, pH, MLR, mordant concentration, dye concentration, and salt concentration) on surface color strength have been evaluated to optimize dyeing conditions. Color fastness to washing, rubbing and light, in general, and dyeing pH sensitivity, in particular, for selective fiber-mordant-dye systems have also been assessed, and it is found that dyeing at pH 11 for the system offers overall good color yield and color fastness properties. Improvement in wash- and lightfastness is also achieved with suitable chemical posttreatment (Samanta et al., 2011).

The combination dyeing using gallnut Al, Cu, Fe sappan wood, and gallnut Al, Cu, Fe gardenia was performed on bast (Jute) fiber of mulberry, cotton, silk, and their K/S values, colors, and sunlight fastness were measured. The gallnut Al, Cu, Fe gardenia dyeing showed the highest K/S values when the dyeing concentration of gallnut is 3%. It tended to show the higher K/S values than gallnut Al, Cu, and Fe sappan wood. The silk showed the highest values of in K/S, and then followed by cotton, and bast fiber of mulberry. The mordants developed different colors on the bast fiber and the cotton treated with gallnut Al, Cu, and Fe sappan wood. However, the silk showed a series of yellow red, showing no effect of the mordants on the development of color. The combination dyeing of gallnut Al, Cu, and Fe gardenia showed a series of yellow. The results showed that sappan wood could develop various colors but gardenia could develop a series of yellow. No distinct improvement on sunlight fastness of the combination dyeing was observed (Park and Yoon, 2011).

6.1.9   Role of anthocyanin of black carrots in fabric dyeing

Anthocyanin can be a good fabric dye as can be understood from their structure (Fig. 6.5) (Shukla and Vankar, 2013) a good dye having many linked rings, e.g., anthroquinone (Nagia and El-Mohamedy, 2007). Being a food material black carrot's anthocyanin can be classified as acid dye (natural pH of dye 6.2) that can play an important role in dyeing fabric especially protein fibers. Anthocyanin being an acid dye very well adheres on silk and gave good shades. Metal mordants in fabric dyeing are suppose to be dye precursors that produce metal complex dyes with the highest lightfastness and wet fastness (http://en.wikipedia.org/wiki/Acid_dye).

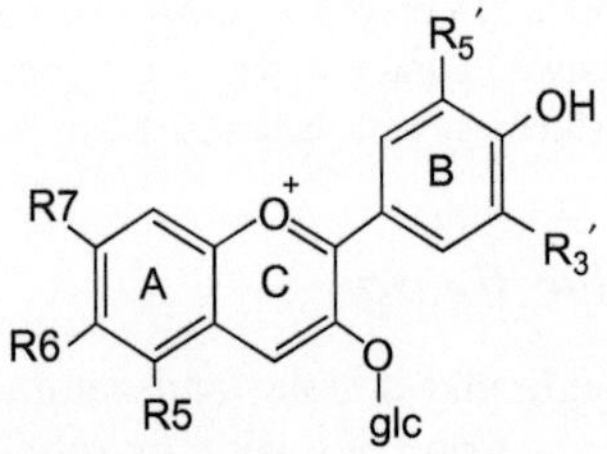

Fig. 6.5 Anthocyanin structure.

6.1.10 Effect of pH in dyeing with Hibiscus extract

Hibiscus extract (Vankar and Shukla, 2011) has a major he color due to anthocyanins. pH has a great effect on anthocyanins. They are redder and more intense in color at low (acid) pH and bluer and less intense in color at a higher pH. This may be observed by the λ_{max} shift from 520 to 552 nm as the pH value increased from 2.55 to 4.5, which was visually confirmed by differences in the color of the solution. At pH 2.0, anthocyanins exist in the colored oxonium or flavylium form (Fig. 6.6A), and at pH 4.5, they are predominantly in the colorless hemiketal form (Fig 6.6B).

The anthocyanin system undergoes a variety of molecular transformations as the pH changes. In slightly acidic aqueous solutions, anthocyanins exist as essentially four molecular species in chemical equilibrium: red flavylium cation, blue quinonialbase,

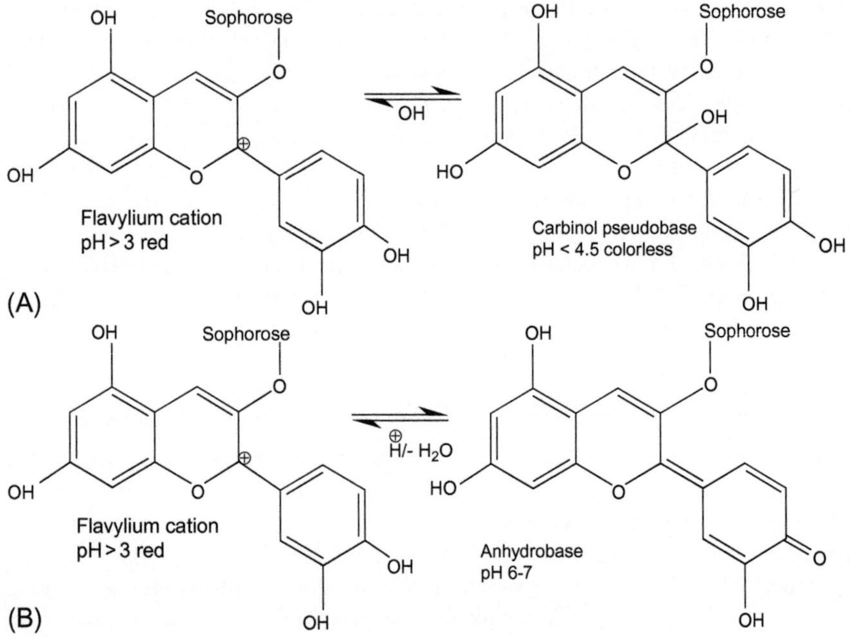

Fig. 6.6 (A) Flavylium form of cyanidin. (B) Hemiketal form of cyanidin.

colorless carbinol pseudobase, and yellowish chalcone. At acidic pH, i.e., 1–3, anthocyanins exist predominantly in the form of the red flavylium cation. Increasing the pH leads to a decrease in the color intensity and the concentration of the flavylium cation that undergoes hydration to produce a colorless carbinol pseudobase. The transmittance of anthocyanin extract of different pH (strong acid to strong alkaline) gave color tones from dark red to mauve. These colors are reproducible and increasing or decreasing pH quickly change color of extract. The λ_{max} values for the anthocyanin extract of *Hibiscus* at different pH are described in Table 6.1.

Table 6.1 **Shift in λ_{max} (nm) in *Hibiscus* anthocyanin with change in pH**

pH	λ_{max}	pH after adding SnCl$_2$	λ_{max}
Intrinsic 2.55	520	1.81	520
3	520	3	538
4	525		
5	545		
6	546	6	560
7	560		
8	578		
9	565	9	552

6.1.11 Common one step and two step dyeing process for dyeing with Delonix extract

A common dyeing was carried for one step dyeing process (simultaneous dyeing) by either using diasterase, lipase, and protease-amylase in 1% w/w with respect to wt of the fabric or biomordant (5%) or metal mordant (2%) in the dye bath as demonstrated. The soured silk fabric was added to this bath along with *Delonix regia* extract (30%, w/w with respect to wt of the fabric).

6.1.11.1 Two step dyeing process for enzyme treated fabrics

Dyeing was carried by a stepwise dyeing process using enzymes such as diasterase, lipase, and protease-amylase, 1% w/w of the fabric (Vankar and Shanker, 2009). The pretreated fabric was used for dyeing with *Delonix regia* extract (30%, w/w with respect to the wt of the fabric).

6.1.11.2 Two step dyeing process for bio-mordanted fabrics

Dyeing was carried by a stepwise dyeing process using 5% biomordant (*Pyrus pashia* fruit extract), w/w with respect to the wt of the fabric. The pretreated fabric was used for dyeing with *Delonix regia* extract (30%, w/w with respect to the wt of the fabric).

6.1.11.3 Two step dyeing process for metal mordanted fabrics

Dyeing was carried by metal mordanting (in the ratio of 1%–2% mordant, w/w with respect to the fabric). The pretreated fabric was used for dyeing with *Delonix regia* extract (30%, w/w with respect to the wt of the fabric).

The dyeing time was 3 h at a temperature of 30–40°C, for all these cases.

6.1.12 Toxicity assay for natural dye (Delonix regia)

Filtrate extract of *Delonix* flowers (1 g/100 mL) was prepared, from this stock different concentrations (25, 50, and 100 ppm) were prepared for testing and finally applied to sterile 9 cm diameter Whatman No. 1 filter paper disks in Petri dishes. Then, 10 surface disinfected green gram (mung beans) were placed on the wetted paper. Ten petri dishes were used as replicates for each treatment. After 7 days of incubation at $27 \pm 1°C$, total root growth (germination) was measured and compared with control (untreated) and expressed as root growth inhibition percentage.

Conclusion

The fabric obtained after weaving is known as gray fabric. It contains both natural and added impurities. In order to make the fabric suitable for dyeing and printing, it is essential to remove the impurities present in gray fabric. The processes involved in the removal of these impurities are known as preparatory processes or fabric pretreatment. The chemical nature of both natural and added impurities present on gray fabric depends on the nature of fiber from which the fabric has been made. For example, the chemical nature of added and natural impurities present on cotton would be different than those present on silk (http://www.handprintingguiderajasthan.in/).

If impurities from fabric are not completely removed before dyeing, following undesirable effects would be obtained:

- Poor water absorbency of fabric due inefficient removal of natural impurities
- Nonuniform absorption of color during dyeing and printing
- Low fastness of color because part of the color would be taken up by size that would be removed during washing of dyed/printed fabric

This would give false impression of low fastness property of color on fabric after dyeing.

Further Reading

- Das, H. and Kalita, D., 2016. Fibers and Dye Yielding Plants of North East India. In *Bioprospecting of Indigenous Bioresources of North-East India* (pp. 77-99). Springer Singapore.
- Kasiri, M. B. and Safapour, S., 2013. Natural dyes and antimicrobials for textiles. In *Green Materials for Energy, Products and Depollution* (pp. 229-286). Springer Netherlands.

- Kumar, J. K. and Sinha, A. K., 2004. Resurgence of natural colourants: a holistic view. *Natural product research, 18*(1), pp. 59-84.
- Rungruangkitkrai, N. and Mongkholrattanasit, R., 2012, July. Eco-Friendly of textiles dyeing and printing with natural dyes. In *RMUTP International Conference: Textiles & Fashion* (Vol. 3, pp. 1-17).

References

Ali, N., El-Mohamedy, R., 2011. Eco-friendly and protective natural dye from red prickly pear (Opuntia Lasiacantha Pfeiffer) plant. J. Saudi Chem. Soc. 15, 257–261.

Ali, N., Mohamedy, R.E., El-Khatib, E., 2011. Antimicrobial activity of wool fabric dyed with natural dyes. Res. J. Text. Appar. 15, 1–10.

Barani, H., Maleki, H., 2011. Plasma and ultrasonic process in dyeing of wool fibers with madder in presence of lecithin. J. Dispers. Sci. Technol. 32, 1191–1199.

El-Zawahry, M., 2009. Ecofriendly dyeing of linen fabric with natural dyes using different enzymes complexes. Man-Made Text. India 52 (10), 337–343.

Ghorpade, B., Darvekar, M., Vankar, P.S., 2000a. Ecofriendly cotton dyeing with Sappan wood dye using ultrasound energy. Colourage 47, 27–30.

Ghorpade, B., Tiwari, V., Vankar, P.S., 2000b. Ultrasound energised dyeing of cotton fabric with Canna Flower extracts using ecofriendly mordants. Asian Text. J. March (3), 68–69.

Gong, Z.-P., Ma, D.-G., 2011. Dyeing behaviour of polyacrylonitrile fiber cured by enzyme for natural dye cochineal. Wool Text. J. 7, 009.

Haji, A., 2010. Functional dyeing of wool with natural dye extracted from *Berberis vulgaris* wood and *Rumex hymenosepolus* root as biomordant. Iran. J. Chem. Chem. Eng. 29, 55–60.

Hegde, M., Bai, S.K., Vijayeendra, M., 2011. Application of Garcinia Indica fruit extract as a mordant for dyeing of organic cotton fabrics for commercial natural colours. Man-Made Text. India 39, 11–16.

Kamel, M., El-Shishtawy, R.M., Yussef, B., Mashaly, H., 2005. Ultrasonic assisted dyeing: III. Dyeing of wool with lac as a natural dye. Dyes Pigments 65, 103–110.

Kamel, M.M., El-Shishtawy, R.M., Youssef, B., Mashaly, H., 2007. Ultrasonic assisted dyeing. IV. Dyeing of cationised cotton with lac natural dye. Dyes Pigments 73, 279–284.

Kamel, M., El Zawahry, M., Ahmed, N., Abdelghaffar, F., 2011. Ultrasonic dyeing of cationized cotton fabric with natural dye. Part 2: cationization of cotton using Quat 188. Ind. Crop. Prod. 34, 1410–1417.

Komboonchoo, S., Bechtold, T., 2009. Natural dyeing of wool and hair with indigo carmine (CI Natural Blue 2), a renewable resource based blue dye. J. Clean. Prod. 17, 1487–1493.

Lee, A.R., Hong, J.-U., Yang, Y.A., Yi, E., 2010. Dyeing properties and antimicrobial activity of silk fabric with extract of unripe Citrus Unshiu fruits. Fibers Polymers 11, 982–988.

Li, H., Zhao, X., 2010. Dyeing performance of natural dyestuff from purple sweet potato to wool fabric. Wool Text. J. 10, 003.

Liakopoulou-Kyriakides, M., Tsatsaroni, E., Laderos, P., Georgiadou, K., 1998. Dyeing of cotton and wool fibres with pigments from *Crocus sativus*—effect of enzymatic treatment. Dyes Pigments 36, 215–221.

Nagia, F., El-Mohamedy, R., 2007. Dyeing of wool with natural anthraquinone dyes from *Fusarium oxysporum*. Dyes Pigments 75, 550–555.

Park, M.-O., Yoon, S.-L., 2011. Properties of natural dyeing of bast fiber (part 3) combination dyeing of gallnut-sappan wood and gardenia. J. Korea Tech. Assoc. Pulp Paper Ind. 43, 1–10.

Parvinzadeh, M., 2007. Effect of proteolytic enzyme on dyeing of wool with madder. Enzym. Microb. Technol. 40, 1719–1722.

Pereira, L., Alves, M., 2012. Dyes—environmental impact and remediation. In: Environmental Protection Strategies for Sustainable Development. Springer International.

Samanta, A.K., Konar, A., Chakraborti, S., Datta, S., 2011. Dyeing of jute fabric with tesu extract: Part 1—Effects of different mordants and dyeing process variables. Indian J. Fibre Text Res. 36, 63–73.

Shanker, R., Vankar, P.S., 2005. Dyeing with *Celosia cristata* flower on modified pretreated wool. Colourage 52, 53.

Shanker, R., Vankar, P.S., 2006. Dyeing silk, wool and cotton with *Alcea rosea* (Hollyhock) flower. Fibre2Fashion.com.

Shukla, D., Vankar, P.S., 2013. Natural dyeing with black carrot: new source for newer shades on silk. J. Nat. Fibers 10, 207–218.

Singh, H.B., Bharati, K.A., 2014. Handbook of Natural Dyes and Pigments. Woodhead Publishing India, New Delhi.

Tiwari, V., Vankar, P.S., 2001. Ecofriendly microwave and sonicator dyeing with natural dyes for Hosiery material. Asian Text. J. 8 (August), 82–85.

Tiwari, V., Ghorpade, B., Vankar, P., 2000a. Dyeing Terrycot and cotton fabric with Lac Dye in Sonicator. Asian Text. J. 9, 68–70.

Tiwari, V., Ghorpade, B., Vankar, P.S., 2000b. Dyeing with aqueous extract of Bougainvillea. Asian Text. J. - Bombay 9, 28–30.

Tiwari, V., Ghorpade, B., Vankar, P.S., 2000c. Ultrasound dyeing with *Impatiens balsamina* using ecofriendly mordants on cotton. Colourage 3 (March), 21–26.

Tiwari, V., Ghaisas, A., Vankar, P.S., 2001. Improved dyeing of hosiery and cotton fabric by Sonicator with Al root bark. Asian Text. J. 10, 111–112.

Tsatsaroni, E., Liakopoulou-Kyriakides, M., 1995. Effect of enzymatic treatment on the dyeing of cotton and wool fibres with natural dyes. Dyes Pigments 29, 203–209.

Vankar, P.S., Shanker, R., 2008. Ecofriendly ultrasonic natural dyeing of cotton fabric with enzyme pretreatments. Desalination 230, 62–69.

Vankar, P.S., Shanker, R., 2009. Potential of *Delonix regia* as new crop for natural dyes for silk dyeing. Color. Technol. 125, 155–160.

Vankar, P.S., Shukla, D., 2011. Natural dyeing with anthocyanins from *Hibiscus rosa sinensis* flowers. J. Appl. Polym. Sci. 122, 3361–3368.

Vankar, P.S., Tiwari, V., Ghorpade, B., 2001a. Microwave and sonicator dyeing of cotton fabric with a mixture of natural dyes using metallic mordant and biomordants. Asian Text. J. 1 (January), 70–73.

Vankar, P.S., Tiwari, V., Ghorpade, B., 2001b. Microwave dyeing of cotton fabric-*Cosmos sulphureus* and comparison with sonicator dyeing. Can. Text. J. 11/12 (November/December), 1–33.

Vankar, P., Tiwari, V., Shanker, R., Singh, S., 2004. *Carthamus tintorius* (Safflower), a commercially viable dye for textile. Asian Dyers 1, 25–29.

Vankar, P.S., Shanker, R., Mahanta, D., Tiwari, S.C., 2007a. Characterization of the colorants from leaves of *Bischofia javanica*. Int. Dyer 193, 31–37.

Vankar, P.S., Shanker, R., Srivastava, J., 2007b. Ultrasonic dyeing of cotton fabric with aqueous extract of *Eclipta alba*. Dyes Pigments 72, 33–37.

Vankar, P.S., Shanker, R., Verma, A., 2007c. Enzymatic natural dyeing of cotton and silk fabrics without metal mordants. J. Clean. Prod. 15, 1441–1450.

Vankar, P., Shanker, R., Dixit, S., Mahanta, D., Tiwari, S., 2008a. Sonicator dyeing of cotton with the leaves extract *Acer pectinatum* Wallich. Pigm. Resin Technol. 37, 308–313.

Vankar, P.S., Shanker, R., Mahanta, D., Tiwari, S., 2008b. Ecofriendly sonicator dyeing of cotton with *Rubia cordifolia* Linn. using biomordant. Dyes Pigments 76, 207–212.

Vankar, P.S., Shanker, R., Dixit, S., Mahanta, D., Tiwari, S.C., 2009. Chemical characterization of the colorants from leaves and stem of Daphne Papyraceae wall and sonicator dyeing of cotton, silk and wool with the plant extract. Pigm. Resin Technol. 38, 181–187.

Zhang, R.-P., Cai, Z.-S., 2011. Study on the natural dyeing of wool modified with enzyme. Fibers Polymers 12, 478–483.

Index

Note: Page numbers followed by *f* indicate figures and *t* indicate tables.